Rafid A Abdulkareem

# Bioethanol production from waste office paper

**Rafid A Abdulkareem**

# Bioethanol production from waste office paper

## by using Microorganisms

**Noor Publishing**

**Imprint**

Any brand names and product names mentioned in this book are subject to trademark, brand or patent protection and are trademarks or registered trademarks of their respective holders. The use of brand names, product names, common names, trade names, product descriptions etc. even without a particular marking in this work is in no way to be construed to mean that such names may be regarded as unrestricted in respect of trademark and brand protection legislation and could thus be used by anyone.

Cover image: www.ingimage.com

Publisher:
Noor Publishing
is a trademark of
International Book Market Service Ltd., member of OmniScriptum Publishing Group
17 Meldrum Street, Beau Bassin 71504, Mauritius
Printed at: see last page
ISBN: 978-620-2-35120-1

Zugl. / Approved by: Baghdad university

# Bioethanol production from waste office paper by using Microorganisms

**By**

**Assistant .Prof Dr .Rafid A. Abdulkareem**

**Osama Fawzi. Saeed**

**Biochemical engineering department .AL Khwarizmi College of Engineering, University of Baghdad. Baghdad, Iraq**

Corresponding authored:      rafid.sigma@yahoo.com

Tele:                           009647901473844

# CHAPTER 1

## INTRODUCTION

## 1.1 BACKGROUND

For the last few years, fossil fuel and its derived products were highly consumed as energy resource for transportation and industrial purposes. Plentiful use of such products results in vast emissions of carbon dioxide ($CO_2$), created due to the combustion of Oil, natural gas and fossil fuels (Saggi and Dey, 2016). The mean temperature on earth has increased by 0.8 °C due to anthropogenic emissions of greenhouse gases (GHG) like the ($CO_2$). At the same time, due to the increase of human population and industrial development, conventional energy sources such as fossil fuels and its derivatives have become unable to meet worldwide energy demand (Saggi and Dey, 2016) .

There are several reasons for the researchers to explore a substitute fuel that is technically viable, environmentally friendly, inexpensive and readily available. First, is the increasing demand for fossil fuel in all fields of human life cycle, be it power generation, manufacturing processes, transport and local consumption (Kafuku and Mbarawa, 2010)(Rajasekaran *et al.*, 2014).

Biofuel is a fuel that is derived from biological materials; it has the potential to be locally and universally available to provide energy. Important area of the modern biotechnology is a conversion of non-edible plant biomass to valuable bio products, biofuels and biochemical (Prema *et al.*, 2015). Biofuel can be produced from biomass (Singh and Trivedi, 2014)(Lu and Zhou, 2014), its mainly include bioethanol, biomethanol, biobutanol, biohydrogen, biodiesel, vegetable oils, biogas, and pyrolysis oils (Demirbas, 2017). Lignocelluloses are the most abundant biomass available on earth; consist mainly of cellulose, hemicellulose and lignin. The most important substrates for production of biofuel among various lignocelluloses are: recycled papers, magazines, newspaper and waste paper (used office paper) (Kuhad *et al.*, 2010).

Ethanol ($C_2H_5OH$) is one of the most common biofuels, which can be used as a replacement for gasoline, other than fuel. Ethanol has an extensive usage as a solvent of materials planned for human connection or consumption, including pharmaceuticals, perfumes, cosmetic, additives and colorings. In chemical industry, ethanol is a crucial solvent and a feedstock for the production of other products. The importance of ethanol is increasing for a several reasons, including the decreasing of global warming and climate change. In the last decade, most research has tended to focus on developing a cost-effective and eco-friendly ethanol production process (Muthsamy *et al.*, 2017).

Bioethanol is a fuel derived from renewable sources, it can be categorized to three main groups: sugary, starchy and lignocellulosic biomass (Srivastava *et al.*, 2015)(Maceiras and Alfonsín, 2017).
First generation (1G) bioethanol is produced from food crops like sugar cane and corn, the plant material used is the edible part of the plant. This is because it can easily be broken down to glucose (Zabed *et al.*, 2016). The

big problem of the (1G) is the compatibility with food supply (Dahnum *et al.*, 2015) (Maceiras and Alfonsín, 2017). Therefore the conversion of edible plant parts into bioethanol is expensive and ethically unacceptable (Azawy *et al.*, 2016).

As a result, a second generation raw materials for bioethanol production are gaining attention, knowing that bioethanol produced from cellulose is exactly the same as that produced from an edible plant material. Cellulosic bioethanol is known as a second generation (2G) bioethanol, and is produced from lignocellulose which is a mixture of lignin, hemicellulose and cellulose while third generation (3G) bioethanol is produced from algae and seaweeds. There is a huge amount of cellulosic waste materials to reuse, waste paper is one of the most important materials that have been used in a large amount, its mainly attractive for producing (2G) bioethanol, since it is rich in carbohydrates and it is readily available. Furthermore, the conversion of waste papers into bioethanol would add a valuable and appreciated method for recycling and handling that waste (Lima *et al.*, 2015) (Adams, 2011).

## 1.2 PROBLEM STATEMENT

Fossil fuel is reducing day by day all over the world. This restriction along with the problem of (GHG) emissions leads discoveries for alternative energy resources that are environmentally and commercially feasible. Every year thousands and thousands tons of waste paper are accumulated that can be used as a feed for biofuel production (Azawy *et al.*, 2016). Compared with the (1G), (2G) biofuel have many benefits; for example, when compared with petrol, most (2G) biofuels are considered to be capable to reduce (GHG) emissions (Sheehan *et al.*, 2004)(Wu, Wang and Huo, 2006). On the other hand, the (2G) biofuel production is challenging, for example acquiring a cheap and stable feed stock, minimizing land use alterations caused by request for biomass feed stock and optimizing bioethanol production technologies. Therefore, waste office paper as a portion of the degradable part in municipal solid waste (MSW) has potential to be a good feed stock for producing bioethanol.

Industrialized bioconversion of lignocellulosic feed stock to bioethanol passes through several stages, where hydrolyzing enzymes are added after pretreatment of the feed stock and finally, microorganisms capable of fermentation are added to produce bioethanol from sugar hydrolysate (Maki, Leung and Qin, 2009). Because of the different usage of paper products their basic structures are variable, which produce variable liabilities for the pretreatment step, which is an essential feature in the economics of the production of bioethanol process since it affects waste treatment, cellulose conversion degrees and hydrolysis of hemicellulose to sugar and their consequent fermentation (Foyle *et al.*, 2007). Bioconversion of waste papers can be enhanced by pretreatments which include physical, chemical and biological pretreatments (Jonsson and Martin, 2016). An effective conversion of waste paper to bioethanol is influenced by the degree of carbohydrate hydrolysis; it can be hydrolyzed using either acid or enzymes. Enzymes are favored over acids because enzymatic hydrolysate is free from any inhibitory products (Byadgi and Kalburgi, 2016). However, the enzymatic hydrolysis of waste papers is delayed by toner's ink, which represents a physical obstacle limiting the hydrolysable positions as well as it release fermentation inhibitory compounds. Amongst several enzymes experienced to deink the waste paper only cellulases were found to be the most active. Cost of cellulase in enzymatic hydrolysis is regarded as a major factor  (Rosenberger *et al.*, 2002).

Meanwhile, industrialized bioconversions of lignocellulosic biomass necessitate multifunctional enzymes with wider substrate consumption as well as the application of enzymes that can work professionally in a wide-ranging of temperature and power of Hydrogen (pH) conditions (Bayer *et al.*, 2004). Therefore, the chance to gain low-priced bioethanol will depend on the positive screening of a novel cellulase producing microorganisms (Sangkharak, Vangsirikul and Janthachat, 2011).

In this study, pretreated waste paper is used to produce bioethanol by using enzymatic hydrolysis process. In general, the current study is important since waste paper is abundant, cheap and represents a promising alternative feed stock for bioethanol production. Further, it overcomes problems related to energy security because bioethanol production from lignocellulosic biomass will substitute the petroleum derived fuel and minimize the (GHG) emissions.


## 1.3 OBJECTIVES

The main objective of this study is to produce bioethanol from pretreated waste office paper using enzymatic hydrolysis. Therefore the following steps are required:

1- Determine the chemical composition of waste office paper collected from local cites and evaluate it as a possible feed stock for bioethanol production.

2- Isolate some types of bacteria from the soil that are capable of decomposition (hydrolysis) of cellulose which can produce cellulase enzyme efficiently.

3- Optimize and investigate the effect of different hydrolysis variables (time, temperature and pH) for maximizing the total reducing sugar (TRS) using statistical methods in Design Expert® Version 7.0 software.

4- Produce bioethanol using separate hydrolysis and fermentation (SHF) process involving bacterial hydrolysis and yeast fermentation after the development of a suitable pretreatment process.

**CHAPTER 2**

**LITERATURE REVIEW**

## 2.1 Bioethanol

Ethanol – also known as ethyl alcohol, alcohol, EtOH, spirit, and ethyl hydroxide with the chemical formula ($C_2H_5OH$), is an important organic chemical because of its exceptional properties, and therefore can be widely used widely for different purposes. In its pure form it is a volatile, flammable, colorless liquid, miscible with water and with many organic solvents and slightly toxic chemical compound with a strong characteristic odor. Table 2.1 shows the chemical and physical properties of ethanol (Balat, 2011) (Anđelković *et al.*, 2017).

The production of ethanol can be done using petrochemical feed stock by the acid-catalyzed hydration of ethylene ($C_2H_4$), or through fermentation process of biomass. Fermentation of sugar into ethanol is one of the earliest organic reactions employed by humanity (Singh and Trivedi, 2014), on a worldwide scale, 75 % of the total production is ethanol produced from petrochemical feed stock, while the rest is produced from biomass fermentation (International Energy Agency, 2015) (Kang *et al.*, 2014).

The primary application of ethanol can be organized into three zones, as a raw material or dissolver in the industry, beverage manufacturing and as fuel. The worldwide production of ethanol as fuel represents about 80 % of the total production in the last few years (Anđelković *et al.*, 2017).

**Table 2.1** Chemical and physical properties of ethanol (Anđelković *et al.*, 2017)

| Molecular formula | $C_2H_5OH$ |
|---|---|
| Molecular weight (g/mole) | 46 |
| Appearance | colorless clear liquid |
| Solubility in water | Fully miscible |
| Azeotrope with $H_2O$ | 95 % ethanol |
| Density (g/cm$^3$) | 0.789 at (25.0 °C) |
| Melting point (°C) | −114.3 |
| Boiling point (°C) | 78.4 |
| Auto ignition temperature (°C) | 363 |
| Viscosity (cp) | 1.2 at (25.0 °C) |
| Heat of evaporation (Btu/lb) | 410 |
| Octane number | 96-115 |

## 2.2 Bioethanol as a fuel

In 1925, Henry Ford an American designer, mentioned ethanol as the fuel of the future. He furthermore stated, "The fuel of the future is going to come from apples, weeds, and saw dust almost anything. There is fuel in every bit of a vegetable matter that can be fermented" (Chandel *et al.*, 2007). Nowadays Henry Ford's innovative visualization meaning can be simply understood. As a transportation fuel, bioethanol has received significant consideration in this field because of its utility as an octane booster, fuel additive, and even as a neat fuel (Banerjee *et al.*, 2010). Bioethanol has a high octane number and a high heat of vaporization which improve ignition, glow-plug, surface ignition and pilot injection are applied to promote self-ignition by using bioethanol blended fuel (Mood *et al.*, 2013) (Krishna and Kakadiya, 2015). It can be used as a total or a partial replacement of gasoline as shown in Table 2.2 (Kang *et al.*, 2014). All vehicles that run on gasoline can use a blend of 10 % ethanol and 90 % gasoline i.e., (E10) without making modifications to their engines. The most general blend for light-duty vehicles is known as E85 and E100, in mixtures such as these, engine modification is necessary (Anđelković *et al.*, 2017).

**Table 2.2** Common ethanol-gasoline mixtures (Kang *et al.*, 2014)

| Code | Composition | Notes |
|------|-------------|-------|
| E5 | Max. 5% EtOH, min. 95% gasoline | Blends for regular cars Engine modification is not needed |
| E10 | Max. 10% EtOH, min. 90% gasoline | |
| E15 | Max. 15% EtOH, min. 85% gasoline | |
| E25 | Max. 25% EtOH, min. 75% gasoline | |
| E85 | Max. 85% EtOH, min. 15% gasoline | Blends for flex-fuel cars Engine modification is needed |
| E100 | Hydrous EtOH | |

For blending with gasoline, purity of 99.5-99.9 % ethanol is required which is called anhydrous ethanol. On the other hand, a mixture that contains 95-96.5 % ethanol (hydrous ethanol) cannot be blended with gasoline in all ratios because it is not miscible with gasoline but can be used as a fuel alone (Niven, 2005). Therefore the water fraction must be removed to burn in combination with gasoline in regular gasoline engines (Prema *et al.*, 2015).

## 2.2.1 Environmental effects of bioethanol fuel

Various development in both industry and transportation over the last decade have helped in increasing the emission of a number of (GHG) such as ($CO_2$), which has caused the deviation in the equilibrium of the earth's atmosphere (Galbe and Zacchi, 2002).

Bioethanol is one of the environment friendly fuels, is comparatively have less effects to the environment, soil and ground water (Aden and Foust, 2009), it is recommended as a biofuel which can be manufactured from renewable feed stock and can decrease the level of (GHG). Using bioethanol is generally ($CO_2$) neutral (Saggi and Dey, 2016) (Saini, 2015), since ($CO_2$) is absorbed by the plant during its growing period and oxygen is released in the same volume of ($CO_2$) produced during bioethanol combustion.

This is a clear advantage of bioethanol over fossil fuel which release ($CO_2$) in addition to other toxic emissions. Figure 2.1 shows the carbon cycle of fossil and biofuel (Mariam *et al.*, 2009) (Krishna and Kakadiya, 2015). Also bioethanol contains 35 % oxygen, which results in fast and complete combustion of fuel and reduces harmful tailpipe emissions. In comparison to conventional gasoline, the blends of E10 resulted in 12-25 % less emission of carbon monoxide (CO) (Zainab and Fakhra, 2014).

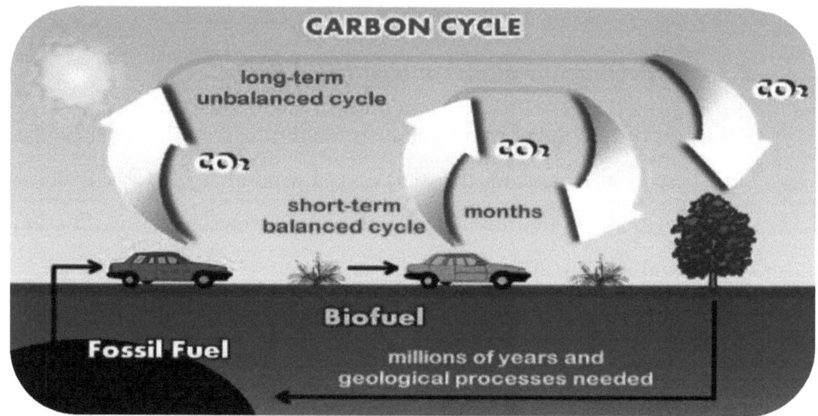


**Figure 2.1** Carbon cycle of fossil fuel vs. biofuel (Krishna and Kakadiya, 2015)

However, volatile organic compounds (VOCs) and nitrogen oxides (NOx) emissions may rise by replacement of gasoline with bioethanol. Nevertheless, bioethanol appears to be a favorable choice when considering the overall pollutants emissions and the net energy gain (Granda *et al.*, 2007).

## 2.3 Raw materials for bioethanol production

Fossil fuel is a non-renewable source of energy and the concern about its sustainability, economic and the environmental impact led to the use of renewable sources for fuel production (Shi *et al.*, 2009). Bioethanol can be produced from different kinds of raw materials, these materials are classified in three main groups: sugary, starchy and lignocellulosic biomass (Srivastava *et al.*, 2015)(Maceiras and Alfonsín, 2017). Lignocellulosic biomass represents an abundant and cheap raw material at reasonable budgets (Balat, 2011).

Biomass resources are often available locally and its processing is possible without high capital investment, energy from biomass is renewable and can contribute to sustainable development. Furthermore, dependence on biomass energy can help in reducing (GHG) emissions (Lynd, 1996).

The various raw materials used in the production of bioethanol via fermentation are classified into three main types: sugars (from sugar cane, sugar beets, molasses and fruits) can be converted to bioethanol directly by fermentation, starches (from corn, cassava, potatoes and root crops) and cellulose (from wood, agricultural residues, waste sulphite liquor from pulp and paper mills) must first be hydrolyzed to fermentable sugars. The sugar and starch based feed stock used in the first and the second generation bioethanol are still used. However, these raw materials may not be enough to meet the increased demand of bioethanol production. Moreover, using human food resources for bioethanol production still a big argument (Zabed *et al.*, 2016) (Taha *et al.*, 2016). Lignocellulosic biomass is available in huge quantities in many countries and this promotes it as an appropriate and potentially inexpensive feed stock for bioethanol production.

### 2.3.1 Lignocellulosic biomass for bioethanol production

Biomass can be described as a natural material generated by living or recently living organisms. Lignocellulose is the term that describes only plants or a plant derived biomass, they are complex mixtures of carbohydrate polymers, namely cellulose, hemicellulose tightly bound to lignin mainly by hydrogen bonds but also by some covalent bonds (Achinas and Euverink, 2016) as shown in Figure 2.2. The main substrates for the production of bioethanol are hemicellulose and cellulose fractions because they are substantial sources of potentially fermentable sugars. Therefore the first step in bioethanol production process is delignification to release cellulose and hemicelluloses from their complex with lignin (Wackett, 2008).

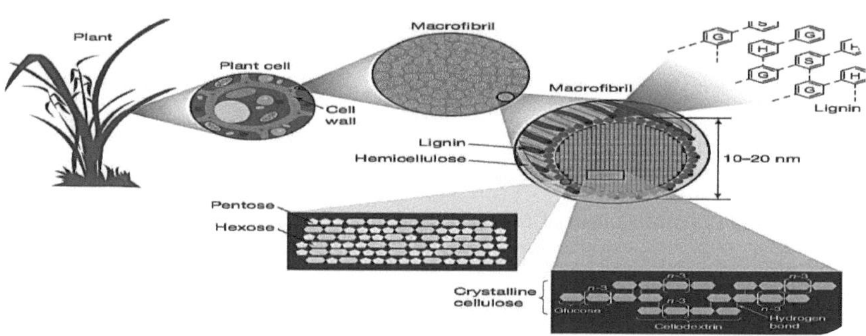

**Figure 2.2** Structure of lignocellulosic feed stock (Achinas and Euverink, 2016)

Lignocellulosic biomass, represent an abundant source for the production of bioethanol. Over other agricultural bioethanol feed stock such as soybeans, sugar cane and corn starch it has the advantage of low-cost, it is a non-edible part of the plant and can be produced at significantly lower price than other food crops (Prema *et al.*, 2015) .

Lignocellulosic biomass is of various types, from grass and soft plants to hardwood and waste paper as shown in Table 2.3. Other sources of lignocellulosic biomass from industrial and agricultural processes include citrus peel waste, sawdust, paper pulp, municipal solid waste and waste paper sludge (Maki, Leung and Qin, 2009).

**Table 2.3** Lignocellulosic biomass and the composition of major components (Silverstein *et al.*, 2007)

| Lignocellulosic material | Cellulose (%) | Hemicellulose (%) | Lignin (%) |
| --- | --- | --- | --- |
| Cotton seed hairs | 80-95 | 5-20 | 0 |
| Corn cobs | 45 | 35 | 15 |
| Coastal Bermuda grass | 25 | 35.7 | 6.4 |
| Grasses | 25-40 | 35-50 | 10-30 |
| Hardwoods stems | 40-55 | 24-40 | 18-25 |
| Leaves | 15-20 | 80-85 | 0 |
| Newspaper | 40-55 | 25-40 | 18-30 |
| Nut shells | 25-30 | 25-30 | 30-40 |
| Paper | 85-99 | 0 | 0-15 |
| Primary wastewater solids | 8-15 | ND | 24-29 |
| Softwoods stems | 45-50 | 25-30 | 30-40 |
| Solid cattle manure | 1.6-4.7 | 1.4-3.3 | 2.7-5.7 |
| Sorted refuse | 60 | 20 | 20 |
| Swine waste | 6.0 | 28 | ND |
| Switch grass | 45 | 31.4 | 12 |
| Wheat straw | 30 | 50 | 15 |
| Waste paper | 60-70 | 10-20 | 5-10 |

NA= Not Define

### 2.3.1.1 Cellulose

Cellulose $(C_6H_{10}O_5)_n$ is the most abundant renewable biopolymer on earth (Pathak and Navneet, 2016) (Bai *et al.*, 2017). It is un branched homogeneous polysaccharide chains each composed of (1000-3000) or more β-D glucose units connected by 1, 4 glycosidic linkage in a linear chain structure. These chains lie parallel to one another allowing for hydrogen bonds and van der Waal's interactions to form between the chains creating mostly crystalline, but also containing amorphous soluble regions, aggregates called microfibrills as shown in Figure 2.3  (Haq , 2016).

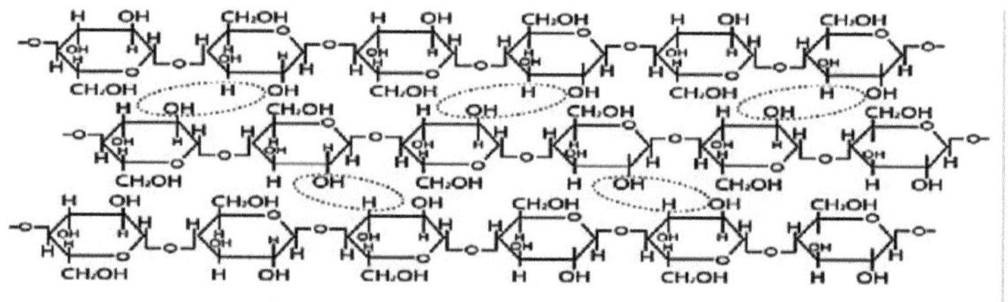

**Figure 2.3** Intermolecular linkages of Cellulose (circles with dotted lines indicate intermolecular hydrogen bonding) (Haq , 2016)

However, cellobiose unit, a dimer of binary glucose is the basic structure block of cellulose as shown in Figure 2.3.1 (Hamelinck *et al.*, 2005).

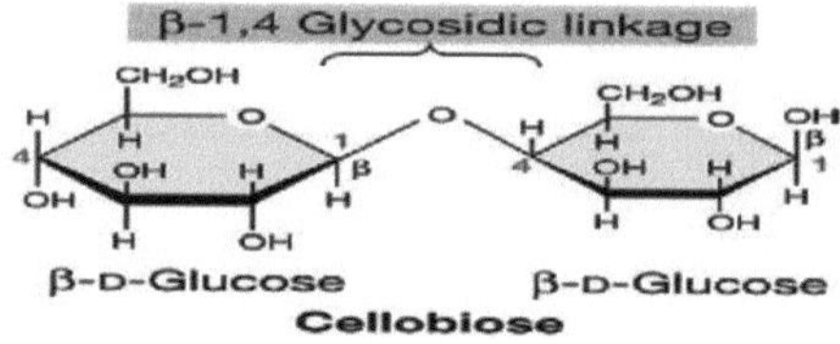


**Figure 2.3.1** Cellobiose unit (Hamelinck *et al.*, 2005)

## 2.3.1.2 Hemicellulose

Hemicellulose $(C_5H_8O_4)_n$ the second major component of lignocellulosic biomass, is a heterogeneous polymer with variable degrees of branching and interaction with other sugar residues. It consists of a five carbon sugar (pentoses) D-xylose, D-arabinose and a six carbon sugar (hexoses) D-mannose, D-glucose, and D-galactose as can be seen in Figure.2.4 (Gírio *et al.*, 2010) (Maki, Leung and Qin, 2009).

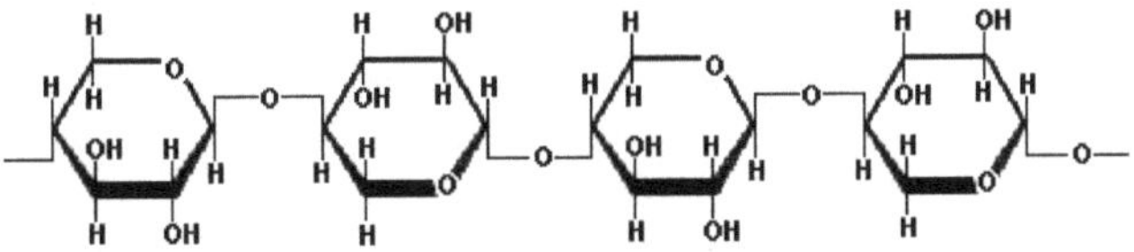

**Figure 2.4** Structure of Hemicellulose (Gírio *et al.*, 2010)

Hemicellulos vary from cellulose by a conformation of several sugar elements and by much smaller and split chains, it has a random, amorphous structure with a slight strength unlike cellulose, which is strong, crystalline and resistant to hydrolysis. Therefore, hemicellulos can be hydrolyzed easily by acids, bases and enzymes (Gírio *et al.*, 2010).

## 2.3.1.3 Lignin

Lignin is the third major components of biomass, is a complex, hydrophobic, aromatic polymer of phenyl propane units that bind covalently to polysaccharides through direct ester and ether linkages. *P*-coumaryl alcohol (I), coniferyl alcohol (II) and sinapyl alcohol (III) are the primary precursors and building units of all lignin as shown in Figure 2.5 (Hamelinck *et al.*, 2005).

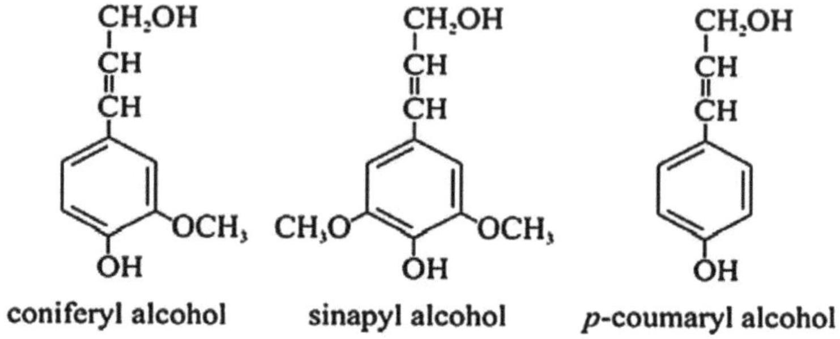

$CH_2OH$ — $CH$ = $CH$ ... (coniferyl alcohol, sinapyl alcohol, p-coumaryl alcohol structures)

**Figure 2.5** Structure of Lignin (Hamelinck *et al.*, 2005)

Because of its structure and heterogeneity, lignin is regarded as the most complex natural polymers, with high resistant to chemical and biological degradation (Lee, 1997).

## 2.4 Waste paper as raw material for bioethanol production

Among lignocellulosic biomass, using waste paper is an exceptional opportunity for bioethanol production due to several reasons; waste paper is an abundant biomass of relatively low cost compared with other biomass materials. It contains low lignin and high carbohydrate content 70-90 % that are potential convertible to bioethanol (Wang *et al.*, 2012a), it is simply digestible without the need to a harsh chemical or physical pretreatment. Thus using waste paper for bioethanol production may offer a useful and an appreciated alternative route to managing this waste in addition to reduce the volume of landfill (MSW) (Dale and Musgrove, 2009).

In our country, Iraq, waste paper which is one of the major constituent of (MSW) has become a severe problem to discard; it is collected and sent to landfill or it is burned. Due to environmental limitations and the absence of suitable new sites, disposal by landfill is becoming expensive, or even impossible, in some areas of the country. Also, burning of this waste is energy-consuming and is environmentally unfriendly because of (GHG) generation and contaminative leachate when the incineration temperature is not high enough (Nishimura *et al.*, 2017).

In recent years, recycling efforts have been supported by legal provisions like the packaging directive. However, when waste papers are reused, it frequently turned into a worse grade paper. Continuous recycling of waste paper, shorten fiber length which impact the paper quality. Therefore the maximum percentage of waste paper recycling is about 65 % (Dubey *et al.*, 2012). This means that a certain fraction of waste paper would always be discarded. This portion contains a substantial and underutilized source of carbohydrate and could be transformed to bioethanol and used for energetic application with environmental benefits. Therefore, waste paper could be used as an outstanding source of lignocellulosic biomass for bioethanol production (Dubey *et al.*, 2012).

## 2.5 Biological processing of lignocellulosic biomass for production of bioethanol

There are several and different steps to produce bioethanol from lignocellulosic feed stock, these steps differ according to the type of the feed stock used. For bioethanol production from lignocellulosic biomass, three major processes are required: pretreatment, hydrolysis and fermentation as shown in Figure 2.6  (Solomon *et al.*, 2007).

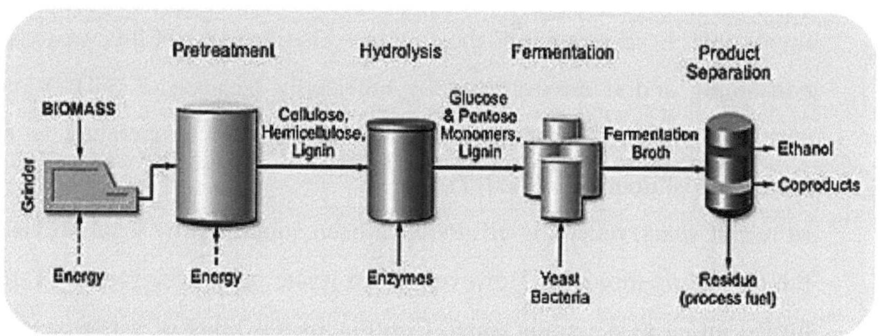


**Figure 2.6** Generalized biomass to bioethanol process (Solomon *et al.*, 2007)

## 2.5.1 Pretreatment

Pretreatment of lignocellulosic biomass is an essential step in bioethanol production process. It involves different processes that change the size, structure and chemical properties of the biomass by disrupting the heterogeneous structure of the cellulosic materials to remove lignin, reduce the crystallinity of the cellulose and to increase the surface area and porosity of the substrate. Thus optimizing the conditions for an efficient hydrolysis, as shown in Figure 2.7 (Mosier *et al.*, 2005).

EFFECT OF PRETREATMENT

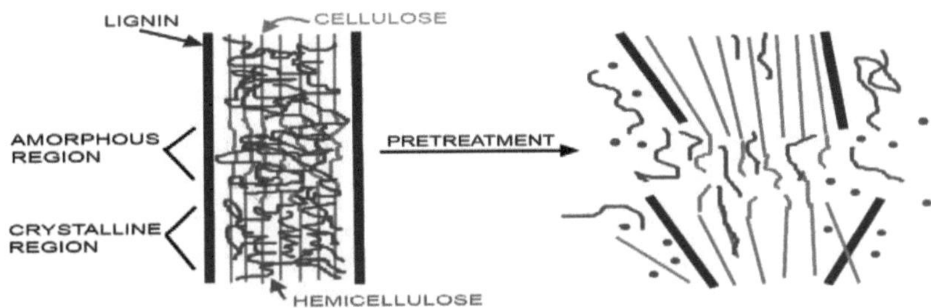


Figure 2.7 Pretreatment effects on biomass (Mosier et al., 2005)

Carbohydrates are firmly bound to lignin mainly by hydrogen bonds and by some covalent connections which make it an intractable feed stock for hydrolysis and bioethanol production. Therefore, pretreatment is a vital stage that breaks the wide connections between cellulose, hemicellulose and lignin components. It makes them more accessible to enzymatic attack before proceeded to hydrolysis and fermentation stages, and can extremely increase the degree of subsequent hydrolysis process (Kumar et al., 2009).

The final products of pretreatment and hydrolysis processes are fermentable sugars (pentose and hexose) and lignin which can be recycled as a fuel to produce heat or electricity (Jonsson and Martin, 2016).

Effective pretreatment must meet the following requirements: increase the sugar production or the capability to consequent form sugars by hydrolysis, evade the carbohydrate degradation, evade the inhibitor compounds formation and cost effective (Haq , 2016).

Since different lignocellulosic biomass have different physiochemical properties, it is necessary to adopt suitable pretreatments technologies based on the properties of used raw materials. Pretreatment methods are generally divided into the following types: physical, physiochemical, chemical and biological methods or a combination of these, each with its own effects on sugar availability and lignocellulosic structure as shown in Table 2.4 (Prasad *et al.*, 2007) (Kumar *et al.*, 2009).

**Table 2.4** Common pretreatment methods for lignocellulosic biomass (Zabed *et al.*, 2016)

| Pretreatment process | Methods | Advantages | Disadvantages |
|---|---|---|---|
| Physical | Mechanical (milling, grinding, extrusion or chipping) Irradiation (electron beam, gamma-ray, microwave, ultrasound). Electric (pulsed electrical field) | No toxic waste and inhibitors from this process. Simple equipment and simple process. | High power consumption. Partial destruction of cellulose, hemicellulose and incomplete disruption of the lignin. Cost depends on the initial and final particle sizes. |
| Physiochemical | Uncatalyzed steam explosion Acid catalyzed steam explosion Auto hydrolysis or liquid hot water (LHW) Ammonia fiber explosion (AFEX) and $(CO_2)$ explosion | Simple, has a short process time. Low enzyme loading is required. | Energy and chemical consuming Requires specific expensive equipment. The recovery of ammonia is necessary to be economically feasible. Expensive and can pollute the environment. Produces inhibitor compounds such as furfural, carboxylic acid. (LHW), (AFEX) are less efficient for biomass with high lignin content. |
| Chemical | Acid and alkaline hydrolysis Ozonolysis Organic solvents Ionic liquids (ILs) | Fast and ideal for biomass with low lignin content. High in delignification and saccharification. | Economic only when solvents are recycled. Expensive and explosive chemicals and solvents. Generation of inhibitory compounds. Water consuming. |
| Biological | Microbes (bacteria, fungi and actinomycetes) particularly brown, white and soft rot fungi | Degrades lignin and hemicelluloses. Environmentally friendly and low energy consumption. Low amount of inhibitory compounds. | Very slow rate of degradation and delignification. Delignification ratio depends on the microbial strains. Cellulolytic activity of the microorganism should be low to reduce the sugar loss. |

### 2.5.1.1 Physical pretreatment

The main objectives of physical pretreatment are: increase specific surface area and pore sizes of biomass, reducing the degree of polymerization of cellulose and the particle size. Physical pretreatments of biomass to bioethanol involves mechanical size reduction, internal delamination and surface fibrillation such as chipping, grinding, milling, extrusion, refining, pyrolysis etc. (Jones *et al.*, 2013). In addition, irradiation by gamma ray, microwave and ultrasound technology have also been evaluated (Nair *et al.*, 2016). Generally, chemicals are not required when doing physical pretreatment. However, most approaches are high energy-demanding with little lignin removal. Therefore, physical pretreatment is combined with chemical pretreatment to lower the overall energy and chemical demand, while still achieving desirable conversions without too much sugar degradation (Chiaramonti *et al.*, 2012).

### 2.5.1.2 Physiochemical pretreatment

Physiochemical pretreatment usually refers to the explosive method to pretreat lignocellulosic biomass. Steam explosion is one of the common and simple methods that involve only water with a reaction temperature ranging from 160 to 260 °C (Sun and Cheng, 2002). This process converts hemicellulose into soluble oligomers with slight impact on lignin degradation (Fernandez-Bolanos *et al.*, 2001), byproduct compounds are generated such as acetic acid, furfural and hydroxyl methyl furfural (HMF) (Gámez *et al.*, 2006). Since these degraded compounds are inhibitory to enzymatic hydrolysis and fermentation process, a   washing step is necessary to remove the inhibitory materials prior to enzymatic hydrolysis (Sun and Cheng, 2002). Auto hydrolysis or liquid hot water (LHW) is another pretreatment method that does not need chemicals. Compared to stream explosion, auto hydrolysis is conducted at a similar temperature range of 130 to 260 °C (Mosier *et al.*, 2005). However, it needs high water demand and does not involve rapid decompression. As a result, higher hemicellulose recovery and lower inhibiting by-products are obtained (Alvira *et al.*, 2010).

Steam explosion with addition of chemicals such as Sulfuric acid ($H_2SO_4$), ($CO_2$) and ammonia fiber explosion (AFEX) can be used to improve pretreatment efficiency (Sun and Cheng, 2002).


### 2.5.1.3 Chemical pretreatment

The chemical pretreatment can be done using several methods includes acid, alkali, Ozonolysis, Organosolv and ionic liquids pretreatment.

The aim of acid pretreatment is to eliminate hemicellulose components of lignocellulosic biomass to enhance hydrolysis step. Inorganic and organic acids such as ($H_2SO_4$), hydrochloric acid (HCl), nitric acid ($HNO_3$), fumaric and maleic acid can be used in the acid pretreatment. Among them ($H_2SO_4$) acid has been most extensively studied, because it is inexpensive and effective. Acid hydrolysis can be divided into concentrated and diluted; a comparison between concentrated and dilute acids hydrolysis is presented in Table 2.5 (Taherzadeh and Karimi, 2007a).

**Table 2.5** Comparison between Concentrated and Diluted Acid Hydrolysis (Taherzadeh and Karimi, 2007a)

| Hydrolysis method | Advantages | Disadvantages |
| --- | --- | --- |
| Concentrated-acid hydrolysis | Operated at low temperature<br>High sugar yield | High acid consumption<br>Equipment corrosion<br>High energy consumption for acid recovery<br>Longer reaction time |
| Diluted-acid hydrolysis | Low acid consumption<br>Short residence time | Operated at high temperature<br>Low sugar yield<br>Equipment corrosion<br>Formation of undesirable by-products |

Concentrated acid, 30 to 70 % at low temperature 40 °C, is effective in terms of breaking down lignocellulosic structure and achieving high sugar conversion (Sivers and Zacchi, 1995). However, using concentrated acid has been avoided due to its corrosive conditions, high cost of equipment and the challenge of acid recovery.

The preferred acid used is ($H_2SO_4$) which provide high hydrolysis yield (Mosier *et al.*, 2005), diluted acid pretreatment using ($H_2SO_4$) followed by enzymatic hydrolysis is found to be an active method for bioethanol production and more favorable for pretreating a wide range of lignocellulosic biomass (Kumar *et al.*, 2009) (Chiaramonti *et al.*, 2012). Dilute acid pretreatment can be performed at low temperature, less than 160 °C for longer periods of time 30 to 60 min; or at high temperature, 160 to 240 °C for shorter periods of time 10 min (Taherzadeh and Karimi, 2007a). The mechanism behind this process is the solubilization of hemicellulose to alter solubilized hemicellulose to fermentable sugars. It is not active in lignin elimination, but can disrupt lignin to a certain extent to increase enzyme accessibility to cellulose (Yang and Wyman, 2004). However, diluted acid pretreatment has several disadvantages. First, the pH of the process stream following pretreatment must be neutralized to the optimum range of enzymes for hydrolysis and further adjustments are necessary for fermentation. This process also produce low sugar yield due to disintegration. Sugar degradation in a single stage diluted acid pretreatment leads to form inhibitory compounds like furfural, (HMF), carboxylic acids, furans and phenolic. In order to decrease sugar degradation, studies have been done using two or more stages of diluted acid hydrolysis with mild conditions in the first stage and more severe conditions in the subsequent stage (Taherzadeh and Karimi, 2007a).

Alkali pretreatment refers to the utilization of base chemicals in the pretreatment process. It is used to remove lignin and some uronic acids which decreases the activity of enzyme (Silverstein *et al.*, 2007). A wide range of alkaline chemicals have been studied for pretreatment use such as: sodium hydroxide (NaOH), calcium hydroxide $Ca(OH)_2$ (lime), potassium hydroxide (KOH) and furthermore  (Alvira *et al.*, 2010). The mechanism behind the alkali pretreatment process is to swell lignocellulosic biomass, causing disruption to lignin structure and linkages between lignin and carbohydrates. An increase of internal surface area, reduction in cellulose crystallinity and degree of polymerization were detected (Sun and Cheng, 2002). To achieve higher lignin removal, alkali pretreatment can be combined with ammonia ($NH_3$) or Hydrogen peroxide ($H_2O_2$) at temperature range of 100 to 220 °C (Martín *et al.*, 2007) (Saha and Cotta, 2007). Alkali pretreatment has shown to cause less sugar degradation than acid pretreatment while requiring a moderate temperature and pressure in comparison to other pretreatment techniques (Alvira *et al.*, 2010). On the other hand, the main disadvantages of alkali pretreatment are that conversion of alkali to irrecoverable salts within the biomass and the need to pH adjustment for subsequent processes (Zabed *et al.*, 2016).

Ozone ($O_3$) pretreatment is very effective in degrading lignin and hemicellulose beside increasing cellulose biodigestibility in a short period of time (Balat, 2011). The main advantages of Ozonolysis are: neither high temperature nor high pressure are used in the Ozonolysis to remove lignin efficiently with no inhibitory compounds formation (Travaini *et al.*, 2013).Thus the process is very well suitable for scaling-up or scaling-down with design simplicity. However, necessitating great quantity of ozone is the major disadvantage of Ozonolysis that makes this process expensive (Kumar *et al.*, 2009).

Organosolv is a pretreatment process using a blend of solvent and acid catalyst to disrupt the hemicellulose and lignin connections. Several organic such as ethanol, methanol, acetone and ethylene glycol are used to solubilize lignin and a part of hemicellulose and provide treated cellulose suitable for enzymatic hydrolysis (Zhao *et al.*, 2009). High temperature (above 185 °C) facilitate delignification to an ideal extent that acid catalysts become unnecessary (Alvira *et al.*, 2010). Organosolv process produces two relatively pure by-products, lignin and an aqueous hemicellulose stream and this is the main advantage of this method (Zhao *et al.*, 2009). However, organosolv process its relatively more expensive due to the high cost of chemical used than steam explosion and the commercial viability highly depends on the complete recovery of the organic solvents (Park *et al.*, 2010).

Ionic liquids (ILs) are the organic salts entirely composed of large organic cations and small inorganic anions, which exist as liquids at wide temperature ranges (usually lower than 100 °C) (Paulová *et al.*, 2013). The unique properties of (ILs) make them an attractive alternative to conventional organic solvents in many chemical processes. It can be used as an effective (green solvents) to break the internal lignin and hemicellulose bonds, due to their excellent thermal and chemical stability, nearly complete non-volatility, non-flammability and negligible vapor pressure (Lee *et al.*, 2010) (Tan *et al.*, 2011). In addition, (ILs) pretreatment is one of the few processes that enable lignin and hemicellulose dissolution and recovery. Recently, 1-butyl-3-methylimidazolium chloride, 1-ethyl-3methylimidazolium acetate and 1-ethyl-3-methyl imidazolium diethyl phosphate have received a wide interest because of their significant cellulose dissolution ability (Lu *et al.*, 2013).

However, discarding of aqueous (ILs) may be environmentally problematic due to their slow degradation and toxicity. Also (ILs) resided on the substrate may negatively affect enzyme and yeast performance. Development of energy efficient recycling methods for (ILs) is a requirement for both economic and technical concern since (ILs) recovery is necessary for cost reduction. Due to the immaturity of this technique and the high cost of the overall operation, this process is still commercially unavailable (Alvira *et al.*, 2010) (Elgharbawy *et al.*, 2016).

### 2.5.1.4 Biological pretreatment

Biological pretreatment is the degradation of lignocellulosic biomass by microorganisms such as bacteria and fungi. It is an environmentally friendly process with several advantages including: no chemical compounds used, lower energy input and the none or low inhibitor compounds production (Keller *et al.*, 2003).

Because of attacking cellulose and hemicellulose, white rot fungi has been found more efficient lignin-degrading agent, while soft and brown rot fungi degrade only cellulose (Matsushika *et al.*, 2009). The main disadvantage of biological pretreatment is the very slow rate of degradation especially in a large scale bioethanol production process (Sun and Cheng, 2002).

### 2.5.2 Hydrolysis (Saccharification)

The lignin removal and hemicellulose hydrolysis are categorized as pretreatment while cellulose degradation is classified as hydrolysis process, and considered as the major hydrolysis step. In hydrolysis, the cellulose is converted into glucose ($C_6H_{12}O_6$) sugars with the help of water as shown in Equation (2.1) (Hamelinck *et al.*, 2005).

$$(C_6H_{10}O_5)n + n\,H_2O \longrightarrow n\,C_6H_{12}O_6 \qquad (2.1)$$

Hydrolysis process can be divided into two types, acid hydrolysis (concentrated or diluted) and enzymatic hydrolysis (Demirbas, 2017). In this process, carbohydrate is broken down to free sugar units; which is called (saccharification). Glucose, a six-carbon sugar or (hexose) is the final product (Haq , 2016).

Acid hydrolysis can be achieved with several types of acids including ($H_2SO_4$), (HCl), ($HNO_3$) and phosphoric acid ($H_3PO_4$). Double acids and a mixture of many acids hydrolysis were also used (Rocha *et al.*, 2011). There are two types of acid hydrolysis, first, concentrated-acid hydrolysis when the biomass treatment performed at high acid concentrations 30 to 70 % and at low temperature, which gives high yields of sugar. The main disadvantages of this process are; high energy and acid consumption, time-consuming (2 to 6 hours) and equipment corrosion (Taherzadeh and Karimi, 2007a) (Banerjee *et al.*, 2010). The second type of hydrolysis is diluted acid which performed at low acid concentrations 0.7 to 3 % and at high temperature 200 to 240 °C. In spite of low acid consumption and shorter time of this process, using the high temperature increases the sugar decomposition rate and equipment corrosion (Taherzadeh and Karimi, 2007a). Decomposition of sugars not only produces a number of inhibitory by-products such as acetic acid, furfural, (HMF) and phenolic compounds, but also lowers the final yield of sugars, Figure 2.8 shows the inhibitory by-products produced from lignocullosic biomass (Klinke *et al.*, 2004).

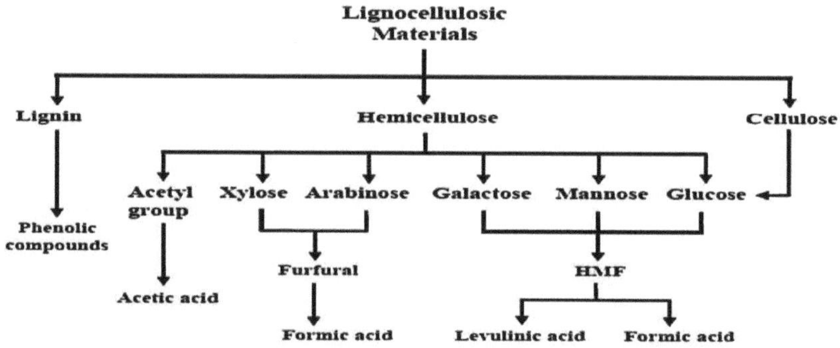


**Figure 2.8** Inhibitory by-products produced from lignocullosic biomass (Klinke *et al.*, 2004)

Enzymatic hydrolysis using enzyme as the catalysts to disrupt the glycosidic linkages, carbohydrates are broken down in enzymatic hydrolysis into reducing sugars that can be fermented to bioethanol by microorganisms (bacteria or yeasts) as shown in Figure 2.9 (Sukumaran *et al.*, 2005).

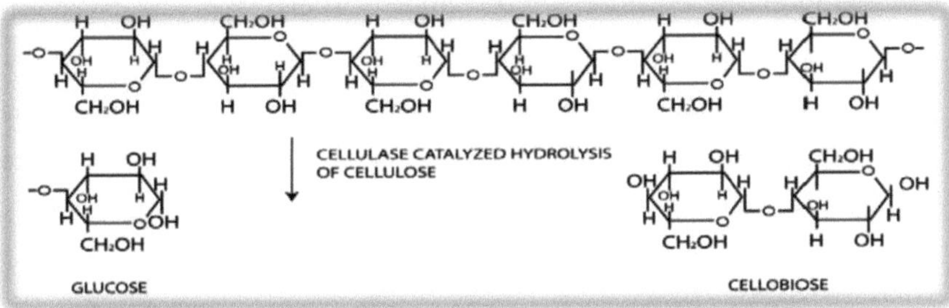


**Figure 2.9** Enzymatic hydrolysis of cellulose (Sukumaran *et al.*, 2005)

Enzymatic hydrolysis is more favorable than acid hydrolysis and considered environmentally friendly. It has higher conversion efficiency, no substrate loss, no production of inhibitory by-products and no need of acid recovery. Also, its more selective and it can be carried out at mild conditions like low temperatures and neutral pH (Harmsen *et al.*, 2010) (Alvira *et al.*, 2010) (Luzzi *et al.*, 2017). Additionally, equipment cost in this process is low compared to alkaline or acid hydrolysis and does not have a corrosion problem (Sun and Cheng, 2002) (Taherzadeh and Karimi, 2007b).

The technical challenges at this step include the high price of enzymes and the slow speed of the process, because the process is hindered by structural parameters of the substrate (cellulose crystallinity, lignin and hemicellulose content and the surface area) (Pan *et al.*, 2006).

Cellulose is hydrolyzed by cellulase, cellulases are enzymes that degrade the $\beta$-1,4-glucan linkages of cellulose microfibrills (Menendez *et al.*, 2015). Cellulase is a complex system of three enzymes that act synergistically to hydrolyze cellulose (Pathak and Navneet, 2016). The optimal temperature and pH for maximum cellulase activity are in the range of 45 to 55 °C and 4 to 6, respectively (Sharada *et al.*, 2014). Cellulase enzyme components are: Endoglucanase (EG)  or carboxymethylcellulase (CMCase), which attacks areas of low crystallinity in the cellulose fiber, Exoglucanase or cellobiohydrolase (CBH), which degrades the particle further by removing cellobiose units from the free chain-ends and $\beta$-glucosidase (BG), which hydrolyzes cellobiose to produce glucose  (Zhang and Zhang, 2013) (Sethi *et al.*, 2013). (BG) in turn, is inhibited by glucose and, therefore, enzymatic hydrolysis is sensitive to the substrate concentration (Andrić *et al.*, 2010). In addition to substrate concentration, enzyme loading, hydrolyzing conditions such as pH, temperature, surfactants and pretreatment method, are the main features affecting the efficiency of enzymatic hydrolysis (Sun and Cheng, 2002) (Sarkar *et al.*, 2012).

Surfactant adsorbing lignin and prevent the unproductive binding between enzyme and lignin and therefore lowers enzyme loading, also an effective pretreatment method is essential to eliminate lignin and to decrease cellulose crystallinity leading to minimize enzyme loading and hydrolysis time (Haq , 2016).

## 2.5.2.1 Microorganisms producing Cellulase

Recently, there are several microorganisms are known as enzyme producers. Both fungi and bacteria can produce cellulase to hydrolyze the lignocellulosic biomass. These organisms are aerobic and anaerobic, mesophilic and thermophilic. Both are widespread and abundant in the nature. Table 2.6 shows most common microorganisms used for cellulase production (Sethi *et al.*, 2013) (Imran *et al.*, 2016) (Nandimath *et al.*, 2016).

**Table 2.6** Common microorganisms used in cellulase production (Otajevwo and Aluyi, 2011) (Sadhu and Maiti, 2013)

| Type | Genus | Species |
|---|---|---|
| Fungi | *Aspergillus*<br><br>*Acremonium*<br>*Fusarium*<br>*Humicola*<br>*Penicillium*<br>*Trichoderma* | *A. niger*<br>*A. fumigatus*<br>*A. Cellulyticus*<br>*F. solani*<br>*H. insolens*<br>*P.fumiculosum*<br>*T. reesai*<br>*T. viride*<br>*T. branchiatum* |
| Bacteria | *Acidothermus*<br>*Bacillus*<br>*Clostridium*<br>*Cytophaga*<br>*Cellulomonas*<br><br>*Pseudomonas* | *A. Cellulyticus*<br>*subtilis*<br>*C. thermocellum*<br>*hutchisonni*<br>*C. uda*<br>*C. fimi*<br>*P. cellulosa*<br>*Flurescens*<br>*aeruginosa* |
| Actinomycetes | *Streptomyces*<br>*Thermonospora* | *S. lividans*<br>*T. fusca* |

Cellulolytic microorganisms are mainly carbohydrate degraders and are generally unable to use proteins or lipids as energy sources for growth (Lynd *et al.*, 2002). Most commonly studied cellulolytic microorganisms are: *Trichoderma, Humicola, Aspergillus, Bacilli, Pseudomonads, Cellulomonas* (Schulein, 1997) (Ong *et al.*, 2004) (Yang *et al.*, 2011) (Chakraborty and Mahajan, 2014) (Nandimath *et al.*, 2016). One of the most extensively studied fungi is *Trichoderma reesei* (Jørgensen *et al.*, 2003) (Ong *et al.*, 2004). The advantages of the cellulase produced by *Trichoderma* are; full complement production of enzyme, resistance to chemical inhibitors and stability under the enzymatic hydrolysis conditions. On the other hand the main drawbacks of *Trichoderma* cellulase are; the low levels and the low activity of (BG) (Taherzadeh and Karimi, 2007b). Recombinant strains of *Trichoderma Sp.* can be used as most creative and influential degrader of cellulose (Balat, 2011).

Cellulases are relatively expensive enzymes; the synthesis cost of cellulase is around 40 % of the total fee in bioethanol production. To reduce this cost, production of crude enzyme on site is viable and cheap due to eliminating the enzyme purification step (Zhang *et al.*, 2006).

There are several reasons for isolation and characterization of newly cellulase from bacteria, these are: the higher growth rate of bacteria, the presence of multi enzyme complexes during synergic action and bacteria can live in a wide diversity of environmental and industrial conditions like alkaliphilic, acidophilic and thermophilic. These conditions produce a very resistive cellulolytic strains to environmental stresses (Sadhu and Maiti, 2013) (Bai *et al.*, 2017). These strains can live and produce enzymes in the severe conditions which were found to be stable under harsh conditions and can be used in the biochemical process (Maki *et al.*, 2009).

Different types of bacteria can be isolated from various environments, and can grow on cellulose and many can catalyze the degradation of soluble derivatives of cellulose. It can be isolated from soil, composite heaps, the feces of ruminants such as cows, decaying plant material from forestry or agricultural waste and other extreme environment like hot-springs (Maki *et al.*, 2009) (Sadhu and Maiti, 2013).

The ability to degrade crystalline cellulose is common for both aerobic and anaerobic bacteria. Aerobic bacteria secrete their cellulases as soluble extracellular enzymes, while anaerobic bacteria are gathered into large complexes called cellulosomes, that attached to the bacterial cell surface (Glazer and Nikaido, 2007). Aerobic bacteria such as *Cellulomonas, Cellovibrio, Pseudomonas* and *Cytophaga* can degrade carbohydrates other than cellulose (Poulsen and Petersen, 1988) (Rajoka and Malik, 1997) (Mawadza *et al.*, 2000).

Anaerobic cellulolytic species have a restricted cellulolytic activity limited to cellulose or its hydrolytic products (Lynd *et al.*, 2002) (Imran *et al.*, 2016). *Cellulomonas Sp.* , Gram-positive, aerobic and rod-shaped bacteria (Glazer and Nikaido, 2007) was isolated from soil  and used as cellulase producing organism (Yin *et al.*, 2010) (Irfan *et al.*, 2012) (Chakraborty and Mahajan, 2014). One of their main distinguishing features is the ability to degrade cellulose, it has the potential to consume lignocellulosic biomass as a carbon source to produce valued enzymes, thus decreasing the cost of enzyme production. The enzyme produced by this strain was thermo stable and alcohol-stable. Therefore it can be used in industrialized processes, especially for bioethanol production (Sangkharak, 2011) (Bai *et al.*, 2017).

### 2.5.3 Fermentation

Fermentation is the bioconversion process of sugars to bioethanol by fermenting microorganisms (yeast or bacteria) which feed with sugar. As the reaction proceeds, sugar is consumed by microorganisms , bioethanol and ($CO_2$) are produced as shown in Equations (2.2) and (2.3) respectively (Krishna and Kakadiya, 2015).

$$3C_5H_{10}O_5 \longrightarrow 5C_2H_5OH + 5CO_2 \qquad (2.2)$$

$$C_6H_{12}O_6 \longrightarrow 2C_2H_5OH + 2CO_2 \qquad (2.3)$$

## 2.5.3.1 Microorganisms used in fermentation process

Microorganisms play a significant role in the production of bioethanol from renewable resources. But when lignocellulosic biomass used as the raw material, the production of bioethanol becomes more challenging and hard. Dissimilar the starchy material feedstock, pretreatment and hydrolysis of lignocellulosic biomass produce a mixture of pentose and hexose sugars along with other inhibiting compounds, which adds difficulties to the fermentation step. Therefore, the important characteristics of the ideal bioethanol-producing microorganisms are, the ability of fermenting both pentose and hexose sugars, minimum nutrient requirements, high bioethanol yield, high tolerance against substrate, bioethanol and inhibitory by-products and no or less byproduct formation (Zyl *et al.*, 2007) (Crespo *et al.*, 2012).

Different microorganisms can be used for bioethanol production. Yeast and bacteria have the ability to ferment carbohydrates under anaerobic conditions (Lynd, 1996). Yeast was the most commonly used microorganism, yeast *S. cerevisiae* can be considered as the most effective microorganism for bioethanol production. It can naturally ferment hexose monosaccharaides (glucose, mannose) and disaccharides (sucrose, maltose) into bioethanol, producing high bioethanol yield shows high tolerance to bioethanol and inhibition by-products (Matsushika *et al.*, 2009) (Najafpour, 2015) (Lima *et al.*, 2015). Furthermore, its genes can be quickly and stably manipulated and it is thus a good candidate to develop for Consolidated Bioprocessing (CBP) technologies.

However, *S. cerevisiae* cannot ferment pentose sugars and this is the main disadvantage that must be considered to overcome (Haan *et al.*, 2013).

Biotechnology-based methods are used to solve such problem; this includes the modification of *S. cerevisiae* that can ferment pentose or by using of another microorganism that naturally ferment pentose sugars (Prema *et al.*, 2015). After the breakdown of the sugars into pyruvate through glycolysis, *S. cerevisiae* converts pyruvate into acetaldehyde through the action of enzyme pyruvate decarboxylase and releases a molecule of ($CO_2$). Acetaldehyde is subsequently reduced by alcohol dehydrogenase to produce bioethanol as shown in Figure 2.10  (Lin and Tanaka, 2006).

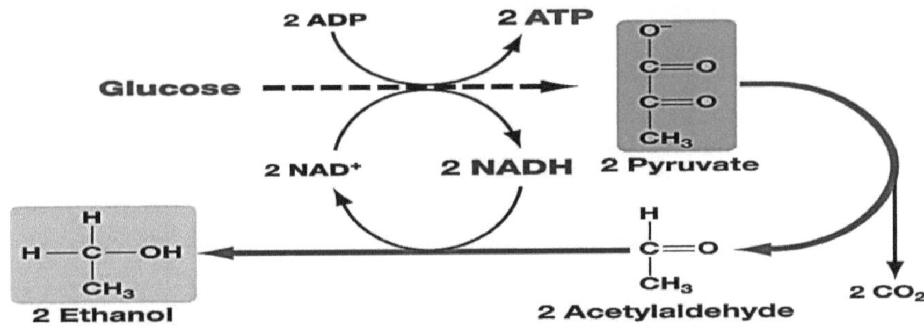


**Figure 2.10** Fermentation steps of glucose  (Lin and Tanaka, 2006)

Compared to yeast, bacteria such as *Pichia stipitis*, *Zymomonas mobilis* and the genetically engineered *Escherichia coli* can produce high bioethanol yield, but they are less resistive to the end product bioethanol, and other compounds present in the hydrolysate (Zhang and Lynd, 2010). The characteristics of the most common microorganisms used for bioethanol production are listed in Table 2.7.

**Table 2.7** Characteristics of most common microorganisms considered for bioethanol production from lignocullosic biomass

| Characteristics | Microorganism | | | |
|---|---|---|---|---|
| | *E. coli* | *P. stipitis* | *S.cerevisiae* | *Z. mobilis* |
| Glucose fermentation | + | + | + | + |
| Other hexose utilization (Galactose and Mannose) | + | + | + | - |
| Pentose utilization (Xylose and Arabinose) | + | + | - | - |
| Direct hemicellulose utilization | - | W | - | - |
| Anaerobic fermentation | + | - | + | + |
| High bioethanol productivity (from glucose) | - | W | + | + |
| Bioethanol tolerance | W | W | + | W |
| Inhibitors tolerance | W | W | + | W |
| Acidic pH range | - | W | + | - |

+=Positive: -=Negative: W=Weak

As can be seen in Table 2.7, *S. cerevisiae* outperforms the other microorganisms in all characteristics except it cannot take up pentose sugars as substrate. From the advantages and disadvantages together, *S. cerevisiae* still remains the prime microorganism for bioethanol production (Bai *et al.*, 2008). Also if the fermentation of xylose is not planned or if xylose concentration is low, *S. cerevisiae* is the chosen microorganism as the best choice for fermenting lignocellulosic hydrolysate (Lima *et al.*, 2015).

## 2.5.3.2 Fermentation and hydrolysis strategies

After the pretreatment and hydrolysis process, the released fermentable sugars become a potential substrate for subsequent fermentation processes. There are a number of strategies for bioethanol production as shown in Figure 2.11 and are described here after.

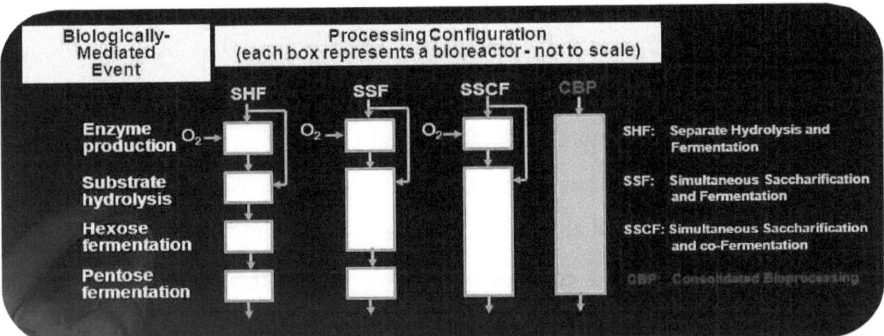


**Figure 2.11** Fermentation and hydrolysis strategies for bioethanol production (Taherzadeh and Karimi, 2007b)

Separate Hydrolysis and Fermentation (SHF) technique refers to the process when the hydrolysis of pretreated lignocelluloses and fermentation processes are done in isolated steps as seen in Figure 2.11 (Taherzadeh and Karimi, 2007b) (Bai *et al.*, 2008) (Dahnum *et al.*, 2015). Due to the separated processes, the major advantage of this method is that the enzymes and fermenting microorganisms can be used at their own optimal conditions temperature and pH (Taherzadeh and Karimi, 2007a). However, the main drawbacks of (SHF) are: enzymes are inhibited by the hydrolysate, and the contamination problems associated with this process. Cellulase is inhibited by the resultant products cellobiose and glucose, and this reduces the efficiency of enzymes and decreases hydrolysis (Zhang *et al.*, 2013). Contamination is another problem in (SHF), the hydrolysis process is long and the diluted solution of sugar provides a suitable environment for microbial contaminations. Enzyme preparation is another source of potential contamination due to difficulties of cellulase sterilization in a large scale process (Taherzadeh and Karimi, 2007b).

In order to reduce the disadvantages of (SHF) technique, Simultaneous Saccharification and Fermentation (SSF) process has been developed where the hydrolysis and fermentation of the hydrolysate are combined in a single step as seen in Figure 2.11 (Saggi and Dey, 2016). In comparison with (SHF), the (SSF) method has higher bioethanol yield, shorter processing time, requires lower amounts of enzyme and lower energy consumption (Sun and Cheng, 2002). Also (SSF) process helps overcoming the inhibition of the cellulase enzyme, because the sugars released (glucose and cellobiose) during hydrolysis  are immediately fermented into bioethanol by the fermenting microorganisms (Dahnum *et al.*, 2015). The presence of bioethanol also helps in eliminating microbial contaminations (Zhang *et al.*, 2013) (Saggi and Dey, 2016). However, the main disadvantage of (SSF) is the difference between optimal temperatures of hydrolyzing enzymes 45 to 55 °C and the fermenting microorganisms 30 to 35 °C (Kádár *et al.*,  2004) (Öhgren *et al.*, 2007). Another approach of fermentation process is Simultaneous Saccharification and Cofermentation (SSCF) approach, where fermentation of both pentose and hexose sugars is performed to produce bioethanol as seen in Figure 2.11 (Kang *et al.*, 2010) (Kang *et al.*, 2014).

In all of the processes discussed, an isolated enzyme production unit is necessary, or the enzymes should be provided externally.

Consolidated Bioprocessing (CBP) of lignocellulosic biomass refers to the combination of all the biological actions required for bioethanol production, i.e. (production of enzymes, hydrolysis of cellulose and hemicellulose present in the pretreated biomass, fermentation of hexose and pentose sugars) in a single unit by a single microorganism's community as seen in Figure 2.11.

The (CBP) process is also known as Direct Microbial Conversion (DMC) (Taherzadeh and Karimi, 2007b). The (CBP) technique needs extremely engineered microorganisms developed for different specific-process properties. Although no natural microorganism exhibits all the features desired for (CBP), a number of microorganisms, both bacteria and fungi, possess some of the desirable properties (Lynd *et al.*, 2005) (Zyl *et al.*, 2007). A comparison between the different strategies for bioethanol production is shown in Table 2.8.

**Table 2.8** A comparative summary on the processes integration in hydrolysis and fermentation (Chandel *et al.*, 2007) (Gírio *et al.*, 2010) (Balat, 2011)

| Process description | Advantages | Disadvantages |
|---|---|---|
| SHF; Hydrolysis and fermentation are done separately | Both hydrolysis and fermentation can be done under their optimal conditions<br>Fermentation can be run in continuous mode with cell recycling | Risk of end product inhibiting the enzymes during hydrolysis<br>Risk for osmotic stress on yeast at high sugar level<br>Risk for contamination |
| SSF; Cellulose hydrolysis and fermentation process are done at the same time | Utilization of high concentration of dry matter<br>End product enzymes inhibitors can be avoided<br>High bioethanol yield<br>Low processing time<br>Requiring low amounts of enzyme<br>Reduced chance of microbial contamination | High viscosity may results in a more power consumption<br>Low mixing and heat transfer efficiency<br>Requiring a compromise between the optimum conditions of the enzymatic reaction and fermenting organism<br>Cells cannot be recycled as they are mixed with the biomass |
| SSCF; Simultaneous saccharification of both cellulose and hemicelluloses and co–fermentation of both glucose and xylose using recombinant strains | Simultaneous conversion of hexose and pentose sugars<br>High solid load<br>Short processing time<br>High bioethanol yield<br>Low production costs | Most commonly used yeast cannot be used<br>Requiring recombinant strains<br>Low rate of cellulose hydrolysis |
| CBP; Bioconversion of cellulose and hemicellulose into bioethanol using engineered microorganisms without adding any exogenous enzymes | Low cost<br>Desired yield<br>Reduced number of reactors<br>Simplification in the operation<br>Reduction in the cost of chemicals<br>Neither operating cost nor capital investments for extracellular enzymes | Instability of the recombinant microorganisms<br>Low bioethanol yield<br>Formation of by-products (acetic acid, glycerol and lactic acid)<br>Limited tolerance of recombinant organism to bioethanol (3.5% w/v) |

## 2.6 Previous studies on the bioethanol production from waste paper

Many researchers have studied the different technologies for bioethanol production from waste paper in terms of pretreatment methods, enzymatic hydrolysis and fermentation conditions. Different paper grades have been examined including: office paper (Sangkharak, 2011) (Lima *et al.*, 2015) (Varotkar *et al.*, 2016) (Maceiras and Alfonsín, 2017) (Nishimura *et al.*, 2017), newspaper (Kuhad *et al.*, 2011) (Ali and Mohd, 2011) (Guerfali *et al.*, 2014) (Subhedar and Gogate, 2015) (Byadgi and Kalburgi, 2016), carton and paperboards (Kádár *et al.*, 2004) (Sangkharak, 2011).

In 1992, Wayman produced bioethanol from waste paper using commercial and crude enzyme without pretreatment using (SSF) technique and shows that the potential for bioethanol production from waste paper is about 200 liter/ton (Wayman, Chen and Doan, 1992). Brooks used a mixed waste office paper as substrate to produce bioethanol using commercial enzyme, acid pretreatment and *Klebsiella oxytoca P2* as a fermenting microorganism, a yield of 0.426 g ethanol/ g substrate was obtained (Brooks and Ingram, 1995). Ballesteros produced bioethanol from recycled-paper using fed-batch (SSF) technique (Ballesteros *et al.*, 2002).

In 2010 Kuhad used newspaper for bioethanol production without pretreatment; (SHF) and fed batch enzymatic saccharification were applied. The batch and fed batch enzymatic hydrolysate when fermented with *S.cerevisiae* produced 5.64g/L and 14.77g/L bioethanol (Kuhad *et al.*, 2010).

After one year, Ali used *Aspergillus niger and S. cerevisiae* for enzymatic hydrolysis and fermentation of waste newspaper, Two methods of saccharification and fermentation i.e. stationary and shaking were adopted. High yield of bioethanol was obtained in stationary fermentation (Ali and

Mohd, 2011). Lima used *S. cerevisiae* and *Spathaspora passalidarum* for fermentation of waste office paper in (SHF) process, the fermentation with *S. passalidarum* resulted in higher bioethanol formation (3.54 g/L) than the fermentation with *S. cerevisiae*, which corresponds to a hypothetical yield of 0.708 g ethanol/g glucose (Lima *et al.*, 2015). At the same year Preeti used ultrasound-assisted enzymatic hydrolysis (commercial enzyme) to produce bioethanol from waste newspaper. Approximately 2.4 times increase in the release of reducing sugar concentration was obtained by the ultrasound-assisted enzymatic hydrolysis approach (Subhedar and Gogate, 2015).

In 2016, Shruti used (SHF) process with *Cytophagahutchisonni* and *S. cerevisiae* microorganisms for bioethanol production from newspaper (Byadgi and Kalburgi, 2016). Recently, Bilal was producing bioethanol using alkali-enzymatic pretreatment and crude enzyme for hydrolysis of old newspaper waste for bioethanol production. Effects of several processing parameters were optimized that led to significantly higher bioethanol production of 17.8 and 20.4 g/L for alkali and enzyme treated substrates, respectively (Bilal *et al.*, 2017). Miroslava used kitchen blender for grinding pretreatment with commercial enzyme for hydrolysis of waste paper to produce bioethanol. The yeast strains *S. cerevisiae* and *Pichia kudriavzevii* were immobilized by entrapment into poly (vinyl alcohol), the immobilized *S. cerevisiae* was a better bioethanol producer in repeated batch fermentations than *P. kudriavzevii* (Zichová *et al.*, 2017).

# CHAPTER 3

## MATERIALS AND METHODS

### 3.1 Materials and methods

The main objective of the current study is to produce bioethanol from pretreated waste office paper using enzymatic hydrolysis and yeast fermentation; therefore the following equipment and chemicals are required:

### 3.1.1 Equipment

The equipment used in the current study is summarized in Table 3.1 with their brand and origin.

**Table 3.1** Equipment used in the current study

| Equipment | Brand | Origin |
|---|---|---|
| Autoclave | K&K | Korea |
| Analytical balance | KERN | Germany |
| Blender | SAMSUNG | Korea |
| Centrifuge | HERMLE | Germany |
| Furnace | HYSC | Korea |
| Hood (Biological cabinet) | Telstar | Spain |
| HPLC | Agilent Technologies | USA |
| HPLC | Knauer | Germany |
| Incubator | memmert | Germany |
| Microscope | Pro.Way | China |
| Magnetic stirrer hot plate | PHOENIX | Germany |
| Milligram laboratory balances | aeADAM | USA |
| Oven | HYSC | Korea |
| pH meter | HANNA | USA |
| Paper shredder | SUNWOOD | China |
| Shaker (incubator) | JSR | Korea |
| Ultra Violet | SHIMADZU | Japan |
| Ultra Sonic | Elmasonic | Germany |
| Vacuum pump | VALEU | Italy |

## 3.1.2 Chemicals

The chemicals used in the current study are summarized in Table 3.2 with their brand and origin.

Table 3.2 Chemicals used in the current study

| Chemical compound | Brand | Origin |
|---|---|---|
| HCl | Scharlau | Spain |
| $H_2SO_4$ | Scharlau | Spain |
| NaOH | Merck | Spain |
| NaCl | Quelab | UK |
| DNS Reagents | Quelab | UK |
| Yeast extracts | Merck | Spain |
| Peptone | Quelab | UK |
| $Mg\ SO_4.7\ H_2O$ | Merck | Spain |
| $KH_2PO_4$ | Merck | Spain |
| $(NH_4)_2SO_4$ | Scharlau | Germany |
| CMC | Himidia | India |
| Agar | Himidia | India |
| Gelatin | Merck | Spain |
| Potassium sodium tartrate | Quelab | UK |

## 3.2  Substrate collection and Preparation

Waste office paper, which was selected as a substrate for bioethanol production, was collected from the Biochemical Engineering Department/ University of Baghdad. The collected waste papers were cut into small size using a standard office shredder; the waste paper size was reduced to about 3 mm. The composition of waste office paper was analyzed for total carbohydrate (cellulose and hemicellulose), moisture, ash and lignin content according to the National Renewable Energy Laboratory (NREL) standard procedures (Sluiter *et al.*, 2012) (Sluiter *et al.*, 2008).

All experiments were conducted twice and the results were the averaged after the two independent repeats. High Performance Liquid

Chromatography (HPLC) with a refractive index detector Agilent Technologies 1260 infinity was used with a Bio-Rad Aminex HPX-87H column for analyzing the carbohydrates. Samples were run at 60 °C and eluted at 0.6 ml/min with 5mM $H_2SO_4$.

## 3.3 Pretreatment of waste office paper

The pretreatment step was performed at (Biotechnology Research Lab. /Faculty of Chemical Engineering/ Babol Noshirvani University of Technology, Iran). Sequences of pretreatment processes were applied before proceeding to the enzymatic hydrolysis of the substrate. The first stage of pretreatment was soaking of waste office paper in distilled water (DW) at room temperature with a ratio of 1:20, and this last for 24 h as shown in Figure 3.1. During this period, the fibers of the paper are loosed and this make the separation of cellulose, hemicelluloses and lignin components from the paper fiber easier (Rajasekaran *et al.*, 2014). Then, the water was filtered and the waste office paper was dried overnight in an oven at 40 °C. After that waste office paper was grinded in an electric blender to form a fluffy wool, this dried substrate was used as a raw material for the next steps (Wu *et al.*, Huang and Shih, 2014) (Bilal *et al.*, 2017).

Delignification was the second stage of pretreatment, it includes ultrasound (US) assisted alkaline treatment for delignification of the waste office paper. Substrate was treated with 0.1N Sodium Hydroxide (NaOH) with ultra-sonication using a laboratory scale ultrasonic 37 kHz and 280 W, as shown in Figure 3.2. Conditions for the (US) pretreatment of waste office paper were optimized by varying substrate loading 1, 2, 3, 4 and 5

% (w/v), pretreatment time 10, 15, 20, 25 and 30 min and pretreatment temperature 30, 40, 50, 60 and 70 °C. The resultant, pretreated liquid hydrolysate was analyzed for total reducing sugars (TRS) using dinitrosalicylic acid (DNS) method (Miller, 1959). The Procedures for (DNS) test includes:

1- 0.1 to 1 ml sample + 2.9 to 2 ml (DW) + 3 ml (DNS) reagent

2- Heating the mixture at 90° C for 5 to 15 minutes in water bath, to develop the red-brown color.

3- Add 1 ml of a 40 % potassium sodium tartrate (Rochelle salt) solution to stabilize the color.

4- After cooling to room temperature in a cold water bath, record the absorbance with a spectrophotometer at 540 nm.

(a)        (b)

**Figure 3.1** Soaking step (a) before pretreatment (b) after soaking

**Figure 3.2** Delignification step

## 3.4 Enzymatic hydrolysis

### 3.4.1 Isolation of cellulase producing bacteria from soil

A cellulase producing bacteria was isolated from wet and damp local soil 15 to 20 cm under the soil surface; samples were collected from different locations of University of Baghdad for microorganism isolation. The carboxymethyl cellulose (CMC) medium was used as the individual carbon source for the isolation of bacteria. The (CMC) medium contains the following (Irfan *et al.*, 2012):

CMC 10 (g/L)

Magnesium sulfate heptahydrate ($MgSO_4.7H_2O$) 0.25 (g/L)

Dibasic potassium phosphate ($K_2HPO_4$) 2 (g/L)

Peptone 10 (g/L)

Ammonium sulfate $(NH_4)_2SO_4$ 2.5 (g/L)

Agar 10 (g/L)

Gelatin 2 (g/L)

Production medium was sterilized by autoclaved at 121 °C for 20 min with pH adjusted to 7. After cooling down at room temperature, the medium was poured in sterile petri plates, for bacterial isolation a pour plate and serial dilution technique were used for the soil suspension and spreading was done. After that, plates were incubated at 30 °C for 48 h to notice the distinct colonies.

To obtain pure bacterial colonies, repeated streaking was used, the purified colonies were conserved at 4 °C for more identification and screening (Irfan *et al.*, 2012) (Bai *et al.*, 2017).

### 3.4.1.1 Screening and Identification of cellulolytic bacteria

The bacterial isolate was screened for cellulase enzyme production; pure cultures of isolated bacteria obtained from previous step were transferred in (CMC) agar plates. The plates were stained for 15 min   with 1 % Congo red after incubation at 30 °C for 48 h and de-stained with 1M Sodium Chloride (NaCl) bacterial colonies were observed along with zone of clearance. White colored colonies, or zones of bacteria, were visible on the plates indicating cellulose hydrolysis (cellulase production) (Irfan *et al.*, 2012) (Chakraborty and Mahajan, 2014).

The bacterial colonies having significant clear zone was sub-cultured and re-streaked on (CMC) medium again to ensure the purity of isolate. The purified bacterium was kept on (CMC) containing agar slant at 4 °C for further study.

The pure cultures of isolated bacteria were sent to the central public health laboratory/Ministry of Health for identification by morphological and biochemical tests (BARROW, 1993) (Cullimore, 2010).

### 3.4.1.2 Inoculum preparation

Pure cultures of isolated bacteria were inoculated in broth medium containing the following (Irfan *et al.*, 2012):

Glucose 10 (g/L)

Peptone 10 (g/L)

$K_2HPO_4$ 2 (g/L)

$MgSO_4.7H_2O$ 0.25 (g/L)

$(NH_4)_2SO_4$ 2.5 (g/L)

Production medium was sterilized by autoclaved at 121 °C for 20 min with pH adjusted to 7. After cooling down at room temperature, the isolated bacteria was inoculated and incubated in a shaking incubator at 30 °C for 24 h with 150 recycle per minute (rpm) shaking speed. After 24 h of incubation period these vegetative cells were used as inoculum source (Irfan *et al.*, 2012).

### 3.4.1.3 Production of enzyme

The spore suspensions of isolated bacteria were used as inoculum for production of cellulase, the medium used for production of enzyme contains the following (Bai *et al.*, 2017):

CMC 10 (g/L)

Peptone 10 (g/L)

K$_2$HPO$_4$ 2 (g/L)

MgSO$_4$.7H$_2$O 0.25 (g/L)

(NH$_4$)$_2$SO$_4$ 2.5 (g/L)

Production medium was sterilized by autoclaved at 121 °C for 20 min with pH adjusted to 7. After cooling down at room temperature, the medium was inoculated with 5 % (v/v) of vegetative cells of isolated bacteria and incubated in a shaking incubator at 30 °C for 48 h with 150 (rpm) shaking speed.

After 48 h of incubation, the broth was centrifuged at 10000 (rpm) at 4 °C for 10 min, the clear supernatant gained was used as crude enzyme (Bai *et al.*, 2017).

### 3.4.1.4 Activity of enzyme

Enzyme activity was measured using (CMC) medium. The reaction mixture contained 0.2 mL of the clear supernatant obtained from the previous step plus 1.8 mL of 1% (CMC) (prepared in 0.05 M citrate buffer at pH 5), the mixture was incubated at 50 °C for 30 min and the reducing sugar liberated was assessed. A plot showing optical density (OD) on standard curve achieved by performing (DNS) reaction on different glucose concentrations is shown in Figure 3.3 (Miller, 1959).

After incubation period the reaction was stopped by adding 3 mL of (DNS) reagent and boiled at 90 °C in a water bath for 10 min. After cooling down at room temperature, the (OD) was measured using ultra violet (UV) spectrophotometer at 540 nm against a blank containing all the reagents minus the supernatant (Sethi *et al.*, 2013). One unit (U) of enzyme activity is defined as the quantity of enzyme required to liberate 1 µmol of glucose per minute under the assay conditions (Varotkar *et al.*, 2016).

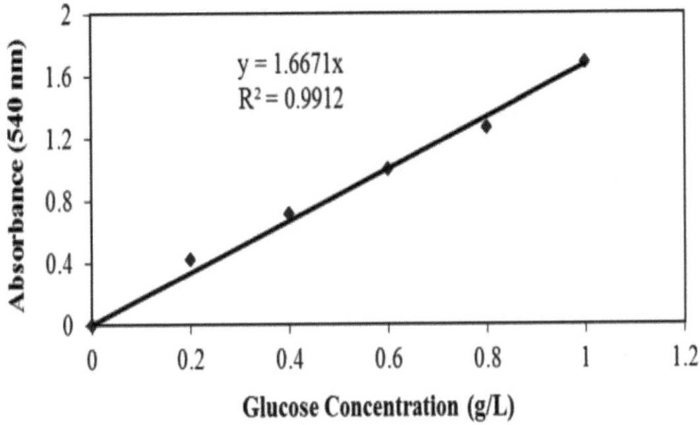


Figure 3.3 Calibration curve of standard glucose concentration

## 3.4.2 Hydrolysis of the pretreated waste office paper
### 3.4.2.1 Hydrolysis process by the conventional method

In the hydrolysis step the cellulose content in the slurry, which produced after pretreatment was converted into reducing sugar utilizing the cellulose degrading bacteria. Enzymatic hydrolysis was done using 5 % (v/v) of supernatant previously prepared at 40 °C with pH adjusted to 5 in an orbital shaker 150 (rpm) for 48 h. Reducing sugars released during hydrolysis were analyzed by (DNS) method every 12 h from zero point.

### 3.4.2.2 Optimization of enzymatic hydrolysis by statistical methods

Optimization of enzymatic hydrolysis was done using Design expert® Version 7.0 software using Response surface methodology (RSM). It is a statistical technique comprising of a collection of mathematical and statistical techniques that can be used to model the response (output variable) of a system influenced by several variables (input variables) (Montgomery, 2013).

Usually, the first stage in the (RSM) is to find an appropriate estimation to the real relationship and the affecting variables between the responses. The most common formulas are first and second order polynomials (Box *et al.*, 1978).

### 3.4.2.3 Design of experiments

To define the best combination of the process variables to optimize the enzymatic hydrolysis of the pretreated waste office paper, the statistical package was used. A second-order central composite design (CCD) was employed, (CCD) is very useful in detecting the significant factors and the interactions between them in a fewer experiments in comparison with the conventional methods (Sangkharak, 2011).

Three factors were selected in this study, the factors selected as independent variables were; $X_1$ (time), $X_2$ (temperature) and $X_3$ (pH). The reducing sugar was considered as the dependent variable (response) (Y). Each factor was evaluated at three coded levels ($-1$, 0 and $+1$). Table 3.3 shows the coded levels and actual values of the three factors.

Plus (+) and minus (−) signs indicate the positive and negative effects on the response. Interaction of the factors, however, had a distinct effect on hydrolysis optimization specifying the importance of these factors for the improvement of the reducing sugar production.

For three variables (n=3), a total of 20 runs were conducted to optimize the hydrolysis process, with (8) factor points, (6) axial points and (6) replicates at the center points. The final experimental design is shown in Table 3.4.

**Table 3.3** Experimental design factors and their levels

| Factor | Unit | Coded levels | | |
|--------|------|----|---|----|
| | | -1 | 0 | +1 |
| Hydrolysis time | h | 24 | 48 | 72 |
| Hydrolysis temp. | °C | 30 | 40 | 50 |
| Hydrolysis pH | | 4 | 5 | 6 |

The second-order polynomial equation was employed to fit the experimental data; the suggested model for the responses (Y) is given in the Equation (3.1) (Box *et al.*, 1978).

$$Y_i = \beta_0 + \beta_1 X_1 + \beta_2 X_2 + \beta_3 X_3 + \beta_{11} X^2_1 + \beta_{22} X^2_2 + \beta_{33} X^2_3 + \beta_{12} X_1 X_2 + \beta_{23} X_2 X_3 + \beta_{13} X_1 X_3 \qquad (3.1)$$

Where ($Y_i$) is the predicted response, ($X_1$, $X_2$ and $X_3$) are independent variables, ($\beta_0$) is the offset term, ($\beta_1$, $\beta_2$ and $\beta_3$) are linear effects, ($\beta_{11}$, $\beta_{22}$ and $\beta_{33}$) are squared effects, and ($\beta_{12}$, $\beta_{23}$ and $\beta_{13}$) are interaction term effects. This polynomial leads to the graphical representation known as (RSM).

**Table 3.4** Experimental design with different conditions

| Run | Type | Factor $X_1$ Time (h) | Factor $X_2$ Temp. (°C) | Factor $X_3$ pH | Response TRS (g/L) |
|---|---|---|---|---|---|
| 1 | Center | 48.00 | 40.00 | 5.00 | |
| 2 | Axial | 48.00 | 40.00 | 6.68 | |
| 3 | Axial | 48.00 | 40.00 | 3.32 | |
| 4 | Center | 48.00 | 40.00 | 5.00 | |
| 5 | Center | 48.00 | 40.00 | 5.00 | |
| 6 | Fact | 24.00 | 30.00 | 4.00 | |
| 7 | Fact | 72.00 | 50.00 | 4.00 | |
| 8 | Center | 48.00 | 40.00 | 5.00 | |
| 9 | Axial | 72.64 | 40.00 | 5.00 | |
| 10 | Fact | 72.00 | 50.00 | 6.00 | |
| 11 | Axial | 48.00 | 56.82 | 5.00 | |
| 12 | Axial | 88.36 | 40.00 | 5.00 | |
| 13 | Fact | 24.00 | 50.00 | 4.00 | |
| 14 | Fact | 72.00 | 30.00 | 6.00 | |
| 15 | Axial | 48.00 | 23.18 | 5.00 | |
| 16 | Fact | 72.00 | 30.00 | 4.00 | |
| 17 | Center | 48.00 | 40.00 | 5.00 | |
| 18 | Fact | 24.00 | 30.00 | 6.00 | |
| 19 | Fact | 24.00 | 50.00 | 6.00 | |
| 20 | Center | 48.00 | 40.00 | 5.00 | |

### 3.4.2.4 Statistical Analysis and Model validation

The data obtained from (RSM) were subjected to analysis of variance (ANOVA) to check for errors and the significance of each parameter. The results were then subjected to a multiple regression analysis to find an empirical model that could relate the response measured to the independent variables. By fitting data from the experiments to the proper model, a regression equation was obtained. The accuracy of the model was specified by the values of the regression coefficient ($R^2$), which is the coefficient of determination and characterization how well the regression model fits the data; it is varies between 0 to 1, where 1 represents a perfect model (Montgomery, 2013). Design expert® Version 7.0 was used to conduct a regression analysis and to plot the response surface graphs.

### 3.5 Fermentation

Bioethanol was produced from glucose by a fermenting microorganism according to the Equation (3.2) (Najafpour, 2015).

$$C_6H_{12}O_6 \ (1 \ g) \longrightarrow 2C_2H_5OH \ (0.511 \ g) + 2CO_2 \ (0.489 \ g) \quad (3.2)$$

Fermentation can be performed as a batch, fed-batch or continuous process. Here the Fermentation Process is an anaerobic batch type.

### 3.5.1 Yeast inoculum preparation

The commercial yeast powder *S. cerevisiae* (saf-instant) was used for fermentation process. The inoculum medium for yeast growth contains the following (Najafpour, 2015):

Glucose 10 (g/L)

Peptone 5 (g/L)

Yeast extract 10 (g/L)

$K_2HPO_4$ 3 (g/L)

Production medium was sterilized by autoclaved at 121 °C for 20 min with pH adjusted to 5. After cooling down at room temperature, 5 (g/L) of yeast powder was added. The medium was then placed in a shaking incubator for 24 h at 35 °C and 150 (rpm) (Najafpour, 2015).

### 3.5.2 Fermentation of waste office paper hydrolysate

A filter-sterilized hydrolysate obtained from optimized enzymatic hydrolysis of waste office paper was used as a substrate for production of bioethanol with no glucose supplementation. Because two microorganisms *Cellulomonas sp.* and *S. cerevisiae* were used in this research with different optimum conditions, (SSF) may not offer the appropriate condition for both which might result in a lower efficiency. Therefore, the method of (SHF) was selected for the current study.

Fermentation was performed in 500 mL flasks containing 300 mL of fermentation medium supplemented with the following:

Yeast extract 3 (g/L)

Peptone 5 (g/L)

Production medium was sterilized by autoclaved at 121 °C for 20 min with pH adjusted to 5 and allowed to cooling down at room temperature. 10 % (v/v) of *S. cerevisiae* prepared in previous step was added to the medium and incubated in a shaking incubator at 35 °C for 72 h and 150 (rpm), experiments were conducted twice and the results were the averaged. During the reaction, samples were taken every 12 h from zero point, to analysis for reducing sugar by (DNS) method (Miller, 1959).

To estimate the produced bioethanol, samples were filtered (0.45 $\mu$m filter) and submitted to an (HPLC) analysis with a refractive index detector. Smart line, Knauer was used with a Eurokat H, Knauer column for analysis of bioethanol. Samples were run at 75 °C and eluted at 0.5 ml/min with 0.01N $H_2SO_4$.

# CHAPTER 4

# RESULTS AND DISCUSSION

## 4.1 Composition of waste office paper

The overall compositions of waste office paper were analyzed. By assuming 100 % conversion, glucose concentration was evaluated as cellulose and that of xylose, arabinose, and manose can be evaluated as hemicellulose (Guerfali *et al.*, 2014). The total carbohydrate (holocellulose) represents cellulose and hemicellulose.

The chemical analysis of waste office paper showed that it consisted of 75.2 % holocellulose. Moisture, ash and lignin content were found to be 4.3 %, 9.7 % and 5.4 % respectively. The concentration of holocellulose as an available source for bioethanol production is in the range of 50 to 73.3 % (Wang *et al.*, 2012b). Therefore the results above indicated that the waste office paper used in this study had high total carbohydrate content 75.2 %. This makes waste office paper a good potential renewable feedstock for bioethanol production. Also, the low lignin content 5.4 % in waste office paper makes it more suitable for reducing sugar production because there will be less physical barrier during the enzymatic hydrolysis and no need for complex and energy intensive pretreatment methods to break the lignocellulosic structure for enzymatic hydrolysis.

The overall compositions of waste office paper obtained in this study were similar to previous studies as can be realized in Table 4.1.

Table 4.1 displays that the range of holocellulose in waste office paper measured in different studies was 68.7 to 83.3 % and that for moisture, ash and lignin were 3.2 to 5.35 %, 7.57 to 13.99 % and 0.93 to 9.53 % respectively. The variation of different studies results are likely due to the different percentage of additives like (kaolin and calcite), binders, ink particles and many unknown organic and inorganic deposits from the paper manufacturing process (Foyle *et al.*, 2007).

On the other hand, the measurement technique can also affect the chemical composition of waste office paper. For this reason, it is essential to choose the appropriate method that permits the minimum loss in carbohydrate content.

**Table 4.1** Compositions of waste office paper

| Waste type | Carbohydrate (%) (Glucan+Xylan) | Moisture (%) | Ash (%) | Lignin (%) | Reference |
|---|---|---|---|---|---|
| Office paper | 77.7 | ND | 11.6 | 0.93 | (Wu *et al.*, 2001) |
| Office paper | 80 | 5.35 | 13.99 | 9.53 | (Foyle et al., 2007) |
| Office paper | 71.29 | 4.9 | 7.69 | 5.9 | (Wang *et al.*, 2012b) |
| Office paper | 83.3 | 3.2 | 9 | 1.2 | (Guerfali *et al.*, 2014) |
| Waste paper | 70 | 3.2 | 8.8 | 9.1 | (Nishimura *et al.*, 2017) |
| Office paper | 75.2 | 4.3 | 9.7 | 5.4 | Current study |

ND=Not Defined

## 4.2 Optimization of pretreatment process

In the current study, the optimization of (US) assisted alkaline treatment was carried out at various substrate loading 1 to 5 % (w/v), pretreatment time 10, 15, 20, 25, and 30 min and pretreatment temperature 30, 40, 50, 60 and 70 °C, with 0.1 N NaOH. The optimum conditions were 4 % substrate loading treatment for 25 min at 60 °C; the effect of different parameters was evaluated by measuring (TRS) released after pretreatment process.

### 4.2.1 Effect of substrate loading

Pretreatment of different substrate loading (1 to 5 % w/v) was carried out with 0.1 N NaOH for 20 min at 50 °C. The graphical representation of the results can be seen in Figure 4.1. The maximum (TRS) obtained was 1.43 (g/L) at 4 % waste office paper concentration. An initial increasing of (TRS) formation was noticed when waste office paper percentage increased from 1 to 4 %. At 5 % concentration the reducing sugar decreased because of mass transfer obstacles due to the increased viscosity of the medium. Therefor a 4 % substrate loading was chosen as an optimum condition for further study.

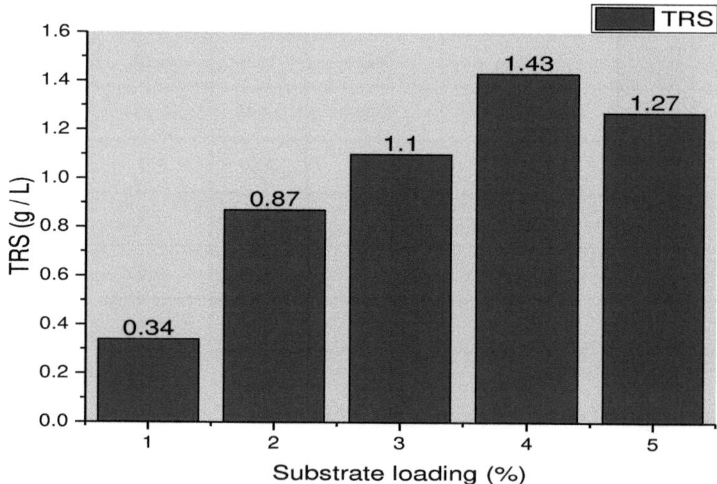


**Figure 4.1** Effect of substrate loading on TRS formation

## 4.2.2 Effect of pretreatment time

Pretreatment of waste office paper was carried out with different pretreatment time 10, 15, 20, 25 and 30 min under optimized substrate loading 4 %, with 0.1 N NaOH at 50 °C. The graphical representation of the results can be seen in Figure 4.2, and the maximum (TRS) obtained was 1.54 (g/L) when the substrate was treated for 25 min.

From the obtaining results, it is observed that there was an increase in the (TRS) content up to 25 min and thereafter there was not much increase, therefore a time of 25 min was chosen as an optimum time for further study.

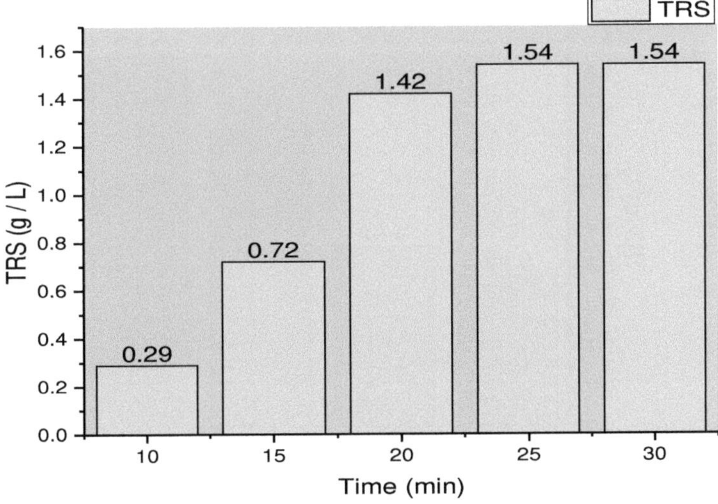


**Figure 4.2** Effect of pretreatment time on TRS formation

### 4.2.3 Effect of pretreatment temperature

Pretreatment of waste office paper was carried out with different pretreatment temperature 30, 40, 50, 60 and 70 °C under optimized substrate loading 4 % and time 25 min with 0.1 N NaOH. The graphical representation of the results can be seen in Figure 4.3, and the maximum (TRS) obtained was 1.89 (g/L) at 60 °C.

An initial increasing of (TRS) formation was observed when the pretreatment temperature increased from 30 to 60 °C, at 70 °C there was a decrease in the concentration of the reducing sugar because of sugar degradation due to high temperature with relatively long time 70 °C for 25 min.

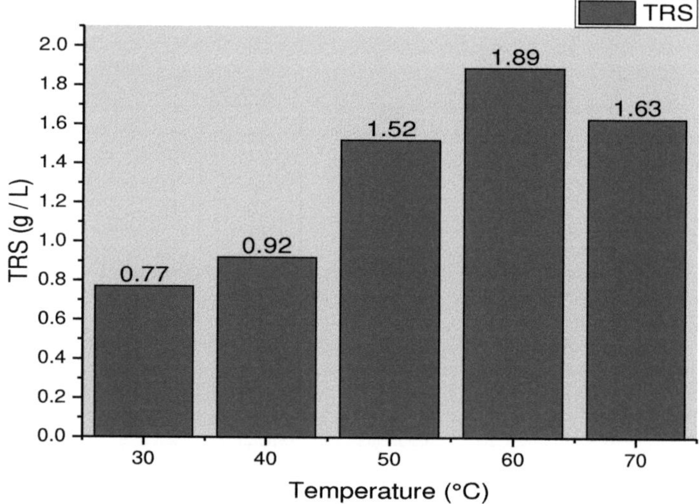


**Figure 4.3** Effect of pretreatment temperature on TRS formation

The results above indicated that (US) assisted alkaline treatment of waste office paper is an efficient pretreatment method for both delignification and hemicellulose degradation for reducing sugar production. US enhances both, the mass transfer rates and the delignification process by increasing the penetration of solvent into the substrate and the generation of turbulence and sound waves  (Subhedar and Gogate, 2015).

## 4.3 Enzymatic hydrolysis

### 4.3.1 Isolation and screening of cellulase producing bacteria from soil

In the current study, a cellulolytic bacterium was isolated from local soil using a pour plate and serial dilution technique. About one third of (DW) and soil samples can produce cellulose degrading bacteria after incubation at 30 °C for 48 h on (CMC) medium under aerobic condition. Other samples did not yield such type of bacteria. Repeated streaking was used and the bacterial colonies were obtained from one third samples as can be seen in Figure 4.4.

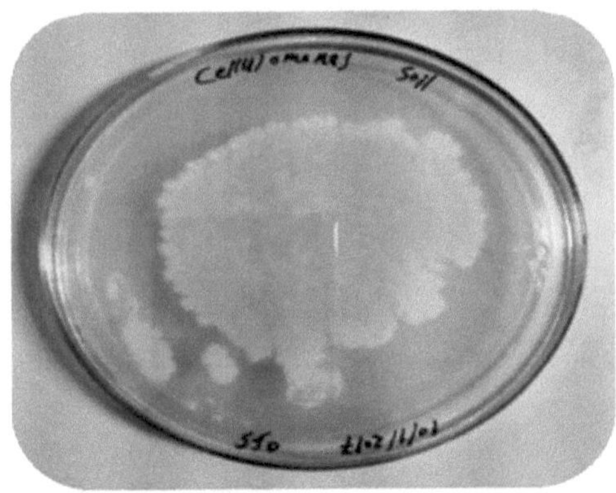

**Figure 4.4** Bacterial colonies

The results shown in Figure 4.5 indicate that, the chosen environment where the cellulosic materials were present was the best source to isolate the cellulolytic bacteria.

These results were similar to previous studies, where researchers have isolated various cellulolytic bacteria from soil (Yin et al., 2010) (Irfan *et al.*, 2012) (Chakraborty and Mahajan, 2014) (Nandimath *et al.*, 2016) (Bai *et al.*, 2017).


## 4.3.2 Screening and Identification of cellulolytic bacteria

The isolated bacterial colonies were screened for cellulase enzyme production; cellulolytic bacteria to be distinguished by significant clear zones in the plates stained with Congo red and de-stained with (NaCl) as shown in Figure 4.5.

These bacterial isolates were sub-cultured and re-streaked on (CMC) medium again to ensure the purity of the isolate. Pure colonies strain was further identified by morphological and biochemical tests.

Microscopic observation of the isolated bacteria shows that it was rod-shaped cells, gram positive upon gram staining. Biochemical reactions and characteristics of the cellulolytic isolated bacteria are summarized in Table 4.2.

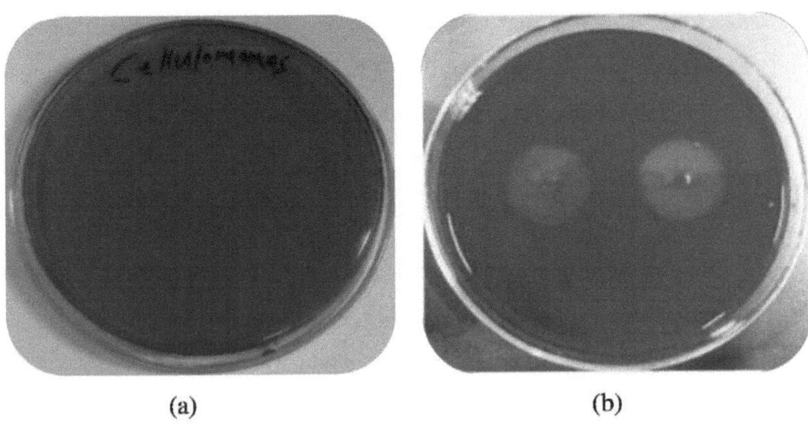

**Figure 4.5** Cellulolytic bacteria growing on (CMC) medium (a) stained with Congo red (b) de-stained with (NaCl)

**Table 4.2** Biochemical reactions and characteristics of the cellulolytic isolated bacteria

| Characteristics /biochemical reaction | Result |
| --- | --- |
| Cell Shape | Rod shape |
| Gram reaction | + |
| Spore formation | - |
| Motility | + |
| Catalase test | + |
| Nitrate reduction test | + |
| Indole test | - |
| Voges-Proskauer test | - |
| Citrate test | - |
| Growth on (CMC) and glucose | + |
| Fermentation of glucose, sucrose and starch | + |

+=Positive: -=Negative

From the results obtained from the morphological and biochemical tests, the bacterial isolate was identified as *Cellulomonas Sp.*

*Cellulomonas Sp.* found to be active producer of cellulolytic enzymes and can represent an attractive, industrial substitute to fungal systems. This is due to their high growth rate, extreme resistance to environmental stress and enzyme complexity  and stability (Otajevwo and Aluyi, 2011) (Sangkharak, 2011) (Sethi *et al.*, 2013) (Bai *et al.*, 2017).

### 4.3.3 Activity of enzyme

Enzyme activity was measured using (CMC) as a substrate according to the procedure explained in section (3.5.1.4), chapter three.
It was found that 16.8 µg of reducing sugar was released, which was inferred from the standard calibration curve of glucose. Therefore cellulase activity was calculated to be 0.35 (U/ml).

### 4.3.4 Hydrolysis of the pretreated waste office paper

The slurry provided to the enzymatic hydrolysis had a (TRS) concentration of approximately 1.89 (g/L) after pretreatment. To explain the (TRS) released only during the enzymatic hydrolysis step, this concentration was subtracted from all the (DNS) results assuming a zero start concentration of reducing sugar prior to enzymatic hydrolysis.
Enzymatic hydrolysis was performed at 40 °C with pH adjusted to 5 for 48 h. The maximum reducing sugars and hydrolysis efficiency were 20.92 (g/L) and 78.4 %, respectively. Figure 4.6 shows the concentration of (TRS) with time.

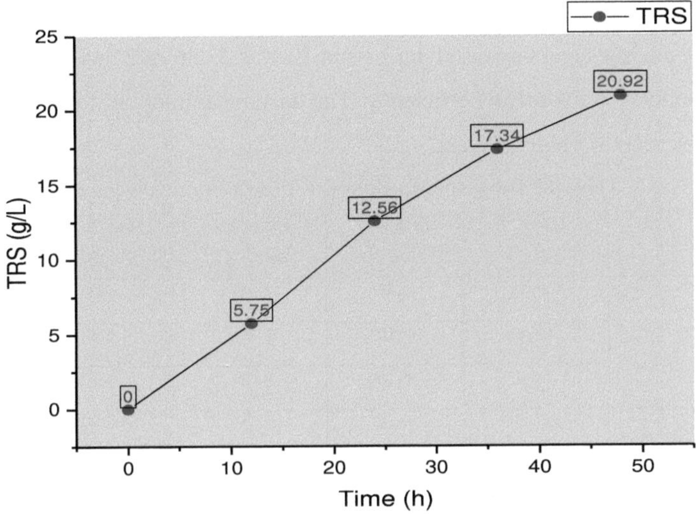


**Figure 4.6** Reducing sugar concentration after enzymatic hydrolysis

In the current study, crude enzymes secreted from bacteria were efficiently utilized for hydrolysis instead of using commercial enzymes, in order to decrease the cost of bioethanol production.

## 4.3.4.1 Optimization of enzymatic hydrolysis by statistical methods

A (RSM) approach was performed to determine the optimal levels of three independent variables; hydrolysis time ($X_1$), hydrolysis temperature ($X_2$) and pH ($X_3$) and the effect of their interaction on the response (Y) of the enzymatic hydrolysis of waste office paper.

Table 4.3 represents the experimental design matrix of the independent variables together with their response to the dependent variable.

The response was found to be in the range of 3.22 to 23.86 g/L, maximum reducing sugars released during run 12 was 23.86 (g/L) which corresponding to 89.4 % hydrolysis efficiency. The design summary of the experiments can be seen in Tables 4.4 and 4.5.

**Table 4.3** Design of experiments with different conditions and responses

| Run | Type | Factor $X_1$ Time (h) | Factor $X_2$ Temp. (°C) | Factor $X_3$ pH | Response TRS (g/L) |
|---|---|---|---|---|---|
| 1 | Center | 48.00 | 40.00 | 5.00 | 21 |
| 2 | Axial | 48.00 | 40.00 | 6.68 | 14.33 |
| 3 | Axial | 48.00 | 40.00 | 3.32 | 12.5 |
| 4 | Center | 48.00 | 40.00 | 5.00 | 20.96 |
| 5 | Center | 48.00 | 40.00 | 5.00 | 20.98 |
| 6 | Fact | 24.00 | 30.00 | 4.00 | 7.87 |
| 7 | Fact | 72.00 | 50.00 | 4.00 | 22.35 |
| 8 | Center | 48.00 | 40.00 | 5.00 | 20.98 |
| 9 | Axial | 72.64 | 40.00 | 5.00 | 23.72 |
| 10 | Fact | 72.00 | 50.00 | 6.00 | 22.97 |
| 11 | Axial | 48.00 | 56.82 | 5.00 | 21.99 |
| 12 | Axial | 88.36 | 40.00 | 5.00 | 23.86 |
| 13 | Fact | 24.00 | 50.00 | 4.00 | 11.42 |
| 14 | Fact | 72.00 | 30.00 | 6.00 | 9.71 |
| 15 | Axial | 48.00 | 23.18 | 5.00 | 3.22 |
| 16 | Fact | 72.00 | 30.00 | 4.00 | 16.31 |
| 17 | Center | 48.00 | 40.00 | 5.00 | 20.98 |
| 18 | Fact | 24.00 | 30.00 | 6.00 | 5.66 |
| 19 | Fact | 24.00 | 50.00 | 6.00 | 13.91 |
| 20 | Center | 48.00 | 40.00 | 5.00 | 20.96 |

The results of the (RSM) experiments were investigated with Design expert® Version 7.0 software. To determine the significant differences of the variables, the experimental data were subjected to an analysis of variance (ANOVA) as listed in Table 4.6.

**Table 4.4** The design summary of (RSM) for independent variables

| Factor | Name | Units | Type | Low Actual | High Actual | Low Coded | High Coded | Mean | Std. Dev. |
|---|---|---|---|---|---|---|---|---|---|
| $X_1$ | Time | h | Numeric | 24.00 | 72.00 | -1.000 | 1.000 | 51.250 | 18.211 |
| $X_2$ | Temp. | °C | Numeric | 30.00 | 50.00 | -1.000 | 1.000 | 40.000 | 8.263 |
| $X_3$ | pH | | Numeric | 4.00 | 6.00 | -1.000 | 1.000 | 5.000 | 0.826 |

**Table 4.5** The design summary of (RSM) for dependent variable (response)

| Response | Name | Units | Analysis | Min. | Max. | Mean | Std. Dev. | Ratio | Model |
|---|---|---|---|---|---|---|---|---|---|
| Y | TRS | g/L | Polynomial | 3.22 | 23.86 | 16.78 | 6.32 | 7.41 | Quadratic |

**Table 4.6** Analysis of variance (ANOVA) for Response Surface Quadratic Model

| Source | Sum of Squares | Mean Square | F Value | p-value (Prob > F) |
|---|---|---|---|---|
| Model | 783.26 | 87.03 | 55.78 | < 0.0001 |
| $X_1$ (Time) | 155.14 | 155.14 | 99.44 | < 0.0001 |
| $X_2$ (Temp.) | 287.56 | 287.56 | 184.32 | < 0.0001 |
| $X_3$ (pH.) | 0.50 | 0.50 | 0.32 | 0.5825 |
| $X_1X_2$ | 7.03 | 7.03 | 4.51 | 0.0597 |
| $X_1X_3$ | 4.90 | 4.90 | 3.14 | 0.1068 |
| $X_2X_3$ | 17.76 | 17.76 | 11.38 | 0.0071 |
| $X_1^2$ | 18.11 | 18.11 | 11.61 | 0.0067 |
| $X_2^2$ | 126.55 | 126.55 | 81.12 | < 0.0001 |
| $X_3^2$ | 103.48 | 103.48 | 66.33 | < 0.0001 |
| Residual | 15.60 | 1.56 | 55.78 | |

The F-value, (calculated by Model Mean Square divided by Residual Mean Square), is a check for comparing model variance with residual (error) variance. If the variances are close to each other, the ratio will be close to unity and it is less possible that any of the factors have a significant effect on the response.

The Model F-value of 55.78 suggests the model significance. There is only a 0.01% chance that a Model F-Value this large could occur due to noise.

The P-value or (Prob > F), is used as a tool to check the significance of each of the coefficients, values of (Prob > F) less than (0.0500) indicate model terms are significant and values greater than (0.1000) indicate the model terms are not significant.

In this case $X_1$ (time), $X_2$ (temp.), $X_2X_3$ (temp.*pH.), $X_1^2$ (time$^2$), $X_2^2$ (temp.$^2$) and $X_3^2$ (pH$^2$) are significant model terms.

The estimated P-values suggest that, independent variables $X_1$ and $X_2$ have a significant effect on the enzymatic hydrolysis of waste office paper; while $X_3$ has no significant effect on the process.

The quadratic term of the three factors, especially in the case of $X_2^2$, $X_3^2$ have a significant effect too. Finally with (P-value<0.0001), the regression model was highly significant and the results show that the model predicted for the degree of hydrolysis of the substrate was suitable.

The quality of the model established was evaluated based on the regression coefficient ($R^2$) and the standard deviation values (Std. Dev.) see Table 4.7.

**Table 4.7** Model adequacy measures

| Std. Dev. | 1.25 | R-Squared | 0.9805 |
|---|---|---|---|
| Mean | 16.78 | Adj. R-Squared | 0.9692 |
| C.V. % | 7.44 | Pred. R-Squared | 0.8082 |
| PRESS | 157.54 | Adeq. Precision | 21.687 |

The Predicted R-Squared of 0.8082 indicates that 80.82% of the total difference in glucose transformation was attributed to the experimental variables studied. However, this may be considered to have an acceptable agreement with the Adjusted R-Squared of 0.9692. The results show a good agreement between experimental and predicted values and infer that the mathematical model is very reliable in predicting the degree of hydrolysis of waste office paper in the current study. Figure 4.7 shows the relationship between the predicted and the actual values of the maximum reducing sugar produced.

Adequacy Precision measures the signal to noise ratio, a ratio greater than 4 is desirable. In this study a ratio of 21.687 indicates an adequate signal.

Finally, the second order polynomial model given in Equation (4.1) was used to determine the degree of waste paper hydrolysis. The same factors, described before were used in this model.

$$Y = 20.99 + 4.10*X_1 + 4.59*X_2 - 0.19*X_3 + 0.94*X_1*X_2 - 0.78*X_1*X_3 + 1.49*X_2*X_3 - 1.46*X_1^2 - 2.99*X_2^2 - 2.70*X_3^2 \qquad (4.1)$$

Where (Y) represents the maximum sugar concentration in (g/L); $X_1$, $X_2$, $X_3$ represent hydrolysis time, hydrolysis temperature and pH respectively. Positive terms have synergetic effect, while negative terms have an antagonistic effect.

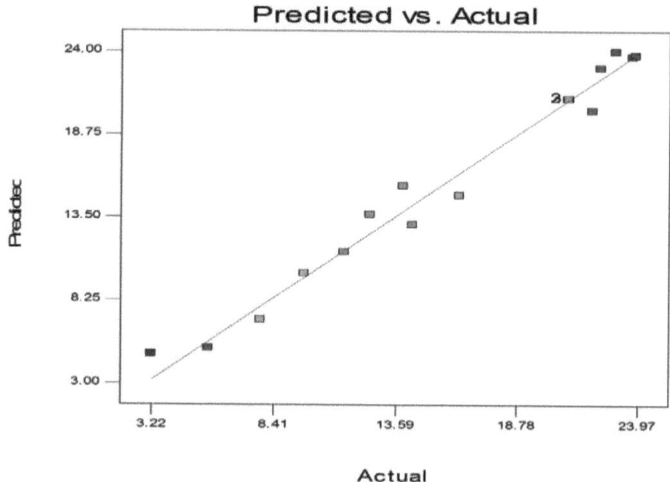


Figure 4.7 Predicted vs. Actual values of maximum reducing sugar produced

## 4.3.4.2 Effect of interactions between independent variables on hydrolysis

The interactions between three variables (hydrolysis time, hydrolysis temperature and pH) on reducing sugar production were investigated using (RSM), and are shown in Figures 4.8, 4.9 and 4.10. The (two-dimensional) contour plots were generated to investigate the combined effect of any two variables on the degree of hydrolysis of waste office paper. This is achieved by varying two variables within certain ranges while keeping the third one constant at central level. The shapes of response surfaces and contour plots indicate the nature and extent of the interaction between different factors, the red region represents the higher reducing sugar produced.

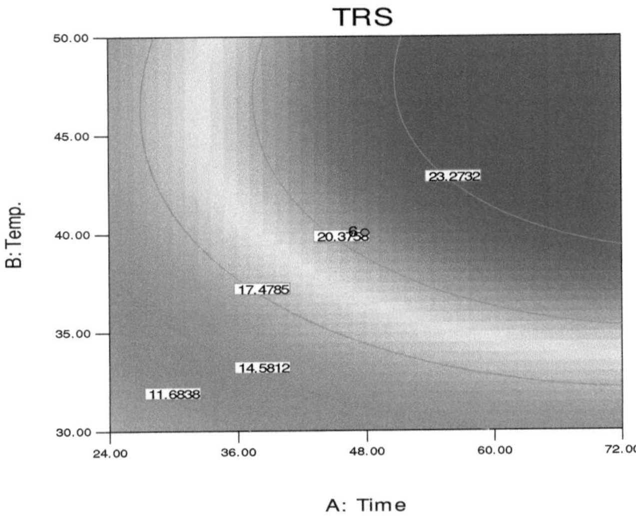


**Figure 4.8** Plot of the combined effects of hydrolysis time and temperature on (TRS) production at constant pH (5)

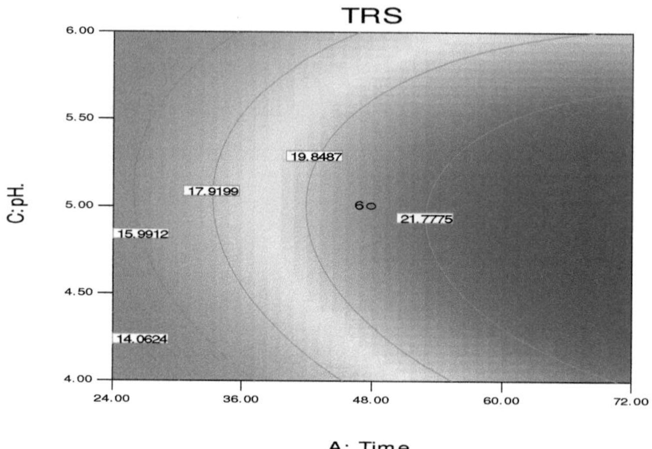


**Figure 4.9** Plot of the combined effects of hydrolysis time and pH on (TRS) production at constant temperature (40 °C)

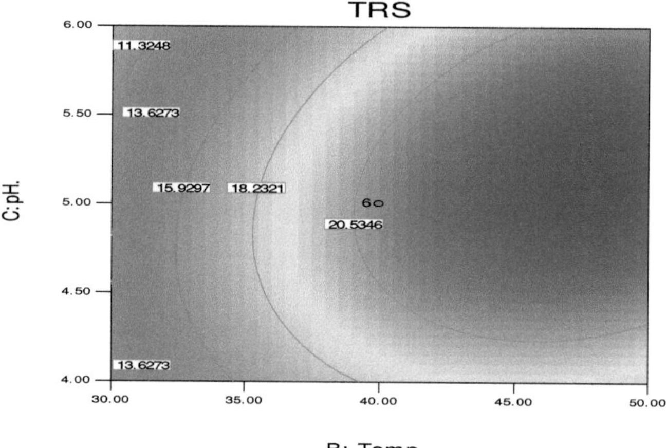


**Figure 4.10** Plots of the combined effects of hydrolysis temperature and pH on (TRS) production at constant time (48 h)

The results from contour plots showed that an interaction between these three parameters had significant effect on sugar production. The predicted response surfaces as functions of hydrolysis time and temperature are presented in Figure 4.8. The (TRS) increased rapidly with increasing hydrolysis time and temperature to optimum conditions. Figure 4.9 shows the predicted response surfaces as functions of hydrolysis time and pH. The (TRS) increased with increasing hydrolysis time and pH to optimum pH 5, and then did not produce a corresponding increasing with a further increase pH. The predicted response surfaces as functions of hydrolysis temperature and pH are presented in Figure 4.10. The (TRS) increased with increasing hydrolysis temperature and pH to optimum conditions 40 °C, pH 5, and then did not produce a corresponding increasing with a further increase.

### 4.3.4.3 Optimization and Model Confirmation

The polynomial model was used to detect optimum conditions for the degree of enzymatic hydrolysis of waste office paper within the experimental ranges investigated. The optimization criteria of hydrolysis of waste office paper are shown in Table 4.8. For three factors, the good process is to maximize the (TRS).

The optimum conditions predicted were 65 h hydrolysis time, 48 °C temperature and pH 4.96, which gave the maximum reducing sugar production 25.537 (g/L), which corresponds to 95.75 % hydrolysis efficiency. The (two-dimensional) contour plots of optimized model are shown in Figure 4.11.

**Table 4.8** Optimization criteria for maximize (TRS)

| Factor | Goal | Lower limit | Upper limit |
| --- | --- | --- | --- |
| Time (h) | is in range | 24 | 72 |
| Temperature (°C) | is in range | 30 | 50 |
| pH | is in range | 4 | 6 |
| TRS (g/L) | maximize | 3.22 | 23.86 |

Using the same optimizing conditions, experimental validation of the model was conducted in replicate. The average value of experimental results was 24.82 (g/L) reducing sugar produced, with 93.1 % hydrolysis efficiency.

The experimental values of reducing sugar can be considered reliable with the predicted values. The convergence relationship between the predicted and the observed values verify the model and the existence of an optimum point. Therefore, the optimum conditions of enzymatic hydrolysis of waste office paper were 65 h time, 48 °C temperature and pH 4.96 which produced 24.82 (g/L) reducing sugar with 93.1 % hydrolysis efficiency.

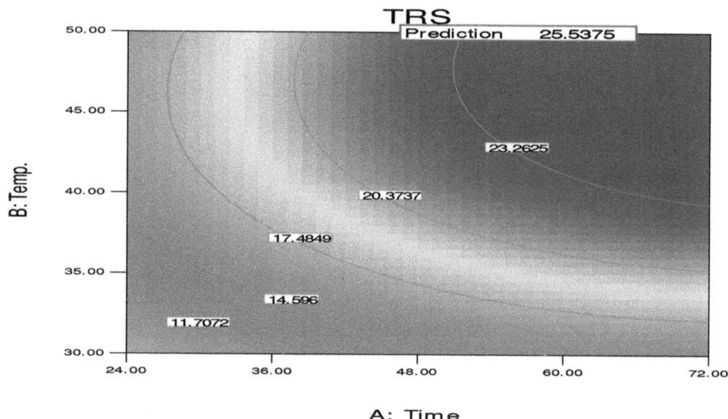


**Figure 4.11** Response surface plots of the optimized model

## 4.4 Fermentation of waste office paper hydrolysate

Fermentation of the optimized hydrolysis hydrolysate which contains 24.82 (g/L) reducing sugar was performed using yeast *S. cerevisiae*.

The maximum concentration of bioethanol estimated using (HPLC) was 11.92 (g/L) after 48 h incubation, the yield and volumetric productivity were 0.498 (g bioethanol/ g glucose) and 0.258 (g bioethanol/ L h) respectively. Conversion efficiency can be found by assuming 1 (g) of glucose will produce 0.511 (g) bioethanol. Hence, conversion percentage of fermentation process would be 93.9 %. The graphical representation of the results of consumption of reducing sugar during fermentation and the produced bioethanol are shown in Figure 4.12.

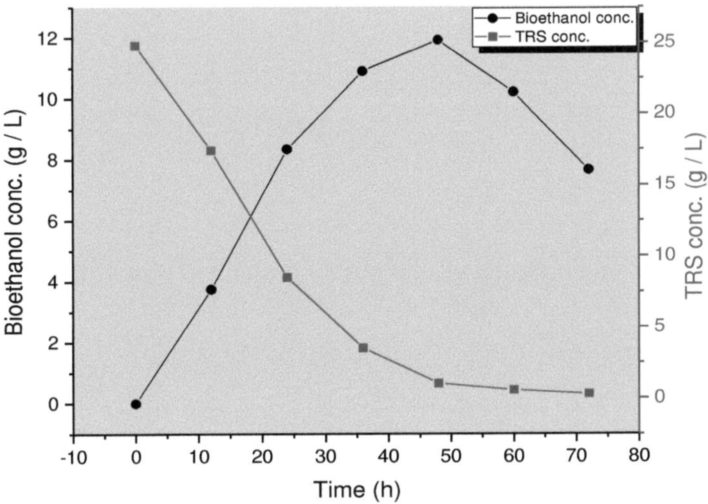


**Figure 4.12** Bioethanol production and sugar consumption during fermentation process

The bioethanol production was found to increase quickly during the first 36 h of fermentation. It starts to decrease only after reaching a maximum concentration of 11.92 (g/L) after 48 h incubation. This indicates the usefulness of extending fermentation time.

The prolonged fermentation process caused reduction in bioethanol concentration after achieving the maximum yield; this may be happened because the yeast *S. cerevisiae* will use accumulated bioethanol as a carbon source for its growth when the concentration of sugar started to decrease. Bioethanol production, substrate type, pretreatment method, hydrolysis method and fermentation process conditions were compared to the previous studies as listed in Table 4.9.

**Table 4.9** Comparison of different bioethanol production process

| Substrate | Pretreatment | Hydrolysis | Fermentation conditions | Bioethanol produced | Reference |
|---|---|---|---|---|---|
| Newspaper | No pretreatment | Enzymatic hydrolysis (commercial enzyme) | (SHF) *S. cerevisiae* | 14.77 g/L (20 h) | (Kuhad *et al.*, 2010) |
| Waste paper | Pretreatment with steam explosion | Enzymatic hydrolysis (crude enzyme) | (SHF) *S. cerevisiae* | 12.5 g/L (48 h) | (Sangkharak *et al.*, 2011) |
| Waste paper | Dilute acid pretreatment | Acid hydrolysis | (SHF) *P. stipitis* | 3.73 g/L (24 h) | (Dubey *et al.*, 2012) |
| Waste newspaper | US- alkaline | US-assisted enzymatic hydrolysis (commercial enzyme) | (SHF) *S. cerevisiae* | 14.1 g/L (48 h) | (Subhedar and Gogate, 2015) |
| Waste office paper | Acid pretreatment | Enzymatic hydrolysis (commercial enzyme) | (SHF) *Spathaspora passalidarum* | 0.42 g/g glucose (24 h) | (Rocha J. M., 2016) |
| Waste paper | Grinding using a kitchen blender | Enzymatic hydrolysis (commercial enzyme) | (SSF) *S. cerevisiae* *P. kudriavzevii* | 17.9 g/L 14.67 g/L (96 h) | (Zichová *et al.*, 2017) |
| Waste office paper | US assisted alkaline | Enzymatic hydrolysis (crude enzyme) | (SHF) *S. cerevisiae* | 11.92 g/L (48 h) | Current study |

It is obvious that the bioethanol yield obtained in the current study was similar to those obtained by previous studies (Sangkharak *et al.*, 2011) (Subhedar and Gogate, 2015) (Zichová *et al.*, 2017). The deviation in the results compared to other studies are likely due to the different pretreatment, hydrolysis and fermentation methods used in the bioethanol production (Rocha , 2016) (Kuhad *et al.*, 2010).

 On the other hand, the microorganisms type used in the hydrolysis and fermentation step can also affect the final bioethanol yield (Ballesteros *et al.*, 2002) (Dubey *et al.*, 2012).

# CHAPTER 5

## CONCLUSION AND RECOMMENDATIONS

## 5.1 CONCLUSION

Because of fossil fuel resources are declining, production of bioethanol from lignocellulosic feedstock has attained importance as a fuel source for the future. This study investigates the possibility of biological conversion of waste office paper to bioethanol. Bioethanol production from waste office paper was accomplished by (US) assisted alkaline pretreatment, enzymatic hydrolysis and fermentation processes.

The conclusions of the present work are listed below as related to the specific objectives set in section (1.3), chapter one:

✓ Characterization of waste office paper

The chemical composition of waste office paper was 75.2 % holocellulose with low lignin content 5.4 %. From this finding it can be concluded that waste office paper is an outstanding lignocellulosic feedstock for bioethanol production.

✓ Isolation of cellulase producing bacteria from soil

A cellulolytic isolate was successfully isolated from local soil and identified as *Cellulomonas Sp*. This typed bacterium has the potential to consume waste office paper as a carbon source to yield cellulase enzyme. A crude enzyme with 0.35 U/ml activity secreted from the isolated bacteria was successfully utilized for cellulose degradation. For industrial bioethanol production, it could be concluded that enzyme production could be enhanced by an optimizing process. Furthermore, by using the crude enzyme instead of the expensive commercial enzyme, the total cost of bioethanol production will be decreased.

✓ Optimization of enzymatic hydrolysis using (RSM)

An optimized enzymatic hydrolysis was done using (RSM) method; the optimization process increases the hydrolysis percentage efficiency from 78.4 to 93.1 %. It can be concluded that using statistical methods at the design stage of the experiments was very helpful to minimize experiments number and to increase the yield and process efficiency.

✓ Bioethanol production using (SHF)

A successful enzymatic hydrolysis and yeast fermentation were performed individually after (US) alkaline pretreatment; bioethanol produced was characterized by (HPLC).

Finally, it can be concluded that production of bioethanol from waste office paper has environmental profits when compared to the landfilling of waste office paper. For example, it reduces the (GHG) emissions from the landfill. Also it can be inferred that consumption of waste office paper for producing biomass based fuel has a positive impact on cost reduction of this promising bioethanol production technology.

## 5.2 RECOMMENDATIONS

For future work several recommendations and new research suggestions can be done based on the finding in the current study:

- Experimentally, waste office paper can be considered as having impressive potential for producing bioethanol. The potential of further advanced technologies in pretreatment, fermentation and other processes, e.g. (SSF) should be evaluated at the laboratory and pilot plant scale.

- Future study could focus on optimization of bacterial isolation and enzyme purification, to achieve maximum availability for the enzymes with higher specific activities and reduce the costs far better than current commercial enzymes. Also, investigations of adding surfactants, substrate loading and enzyme volume and their impacts on the enzymatic hydrolysis must be done.

- Future work should comprise optimization of pretreatment and fermentation process parameters using (RSM). More research is required to improve the performance of each step to obtain the highest yield of bioethanol from waste office paper as a feed stock.

- Yeast *S. cerevisiae* can ferment only hexose sugars into bioethanol, therefore it is recommended that a recombinant yeast or a new combinations of organisms could be combined to change different types of sugars (pentoses and hexoses) into bioethanol for more economical process and higher bioethanol yield.

- Based on the results of the current study on the specific type of paper (waste office paper), it is suggested that further research should be done on bioethanol production from various types of waste paper. For example, magazines, old books and cardboards or even the (MSW) for bioethanol production which would be of interest, from both environmental and economic perspectives.

# References

**Achinas, S. and Euverink, G. J. W.** (2016) 'Consolidated briefing of biochemical ethanol production from lignocellulosic biomass', *Electronic Journal of Biotechnology*. Elsevier B.V., 23, pp. 44–53. doi: 10.1016/j.ejbt.2016.07.006.

**Adams, J.** (2011) 'Analysis of printing and writing papers by using direct analysis in real time mass spectrometry', *International Journal of Mass Spectrometry*. Elsevier B.V., 301(1–3), pp. 109–126. doi: 10.1016/j.ijms.2010.07.025.

**Aden, A. and Foust, T.** (2009) 'Technoeconomic analysis of the dilute sulfuric acid and enzymatic hydrolysis process for the conversion of corn stover to ethanol', *Cellulose*, 16(4), pp. 535–545. doi: 10.1007/s10570-009-9327-8.

**Ali, M. N. and Mohd, M. K.** (2011) 'Production of bioethanol fuel from renewable agrobased cellulosic wastes and waste news papers', *International Journal of Engineering Science and Technology*, 3(2), pp. 884–893.

**Alvira, P., Tomas-Pejo, E. and Ballesterose, M.** (2010) 'Pretreatment technologies for an efficient bioethanol production process based on enzymatic hydrolysis: A review', *Bioresource Technology*. Elsevier Ltd, 101(13), pp. 4851–4861. doi: 10.1016/j.biortech.2009.11.093.

**Anđelković, D. *et al.*** (2017) 'The perspectives of applying ethanol as an alternate fuel', *Energy Sources, Part B: Economics, Planning, and Policy*, 7249(June), pp. 1–10. doi: 10.1080/15567249.2012.683930.

**Andrić, P. *et al.*** (2010) 'Effect and modeling of glucose inhibition and in situ glucose removal during enzymatic hydrolysis of pretreated wheat straw', *Applied Biochemistry and Biotechnology*, 160(1), pp. 280–297. doi: 10.1007/s12010-008-8512-9.

**Azawy, A. N. Al, Khadom, A. A. and Jabbar, A. S. A.** (2016) 'Efficiency of some types of bacteria on producing biofuels from wastes of writing paper', *Journal of Environmental Chemical Engineering*. Elsevier B.V., 4(3), pp. 2816–2819. doi: 10.1016/j.jece.2016.05.033.

**Bai, F. W., Anderson, W. A. and Moo-Young, M.** (2008) 'Ethanol fermentation technologies from sugar and starch feedstocks', *Biotechnology Advances*, 26(1), pp. 89–105. doi: 10.1016/j.biotechadv.2007.09.002.

**Bai, H. *et al.*** (2017) 'Purification and Characterization of Cellulose Degrading Enzyme From Newly Isolated Cellulomonas Sp .', *Cellulose Chemistry and Technology*, 51(3–4), pp. 283–290.

**Balat, M.** (2011) 'Production of bioethanol from lignocellulosic materials via the biochemical pathway: A review', *Energy Conversion and Management*. Elsevier Ltd, 52(2), pp. 858–875. doi: 10.1016/j.enconman.2010.08.013.

**Ballesteros, M. *et al.*** (2002) 'Ethanol production from paper material using a simultaneous saccharification and fermentation system in a fed-batch basis', *World Journal of Microbiology & Biotechnology*, 18(6), pp. 559–561. doi: 10.1023/A:1016378326762.

**Banerjee, S. *et al.*** (2010) 'Commercializing lignocellulosic bioethanol: technology bottlenecks and possible remedies', *Biofuels, Bioproducts and Biorefining*, 4, pp. 77–93. doi: 10.1002/bbb.188.

**BARROW, G. I. and R. K. A. F.** (1993) *Cowan and Steel's manual for the identification of medical bacteria.* 3rd Editio. CAMBRIDGE UNIVERSITY PRESS.

**Bayer, E. A. *et al.*** (2004) 'The Cellulosomes: Multienzyme Machines for Degradation of Plant Cell Wall Polysaccharides', *Annual Review of Microbiology*, 58(1), pp. 521–554. doi: 10.1146/annurev.micro.57.030502.091022.

**Bilal, M. *et al.*** (2017) 'Enhanced Bio-ethanol Production from Old Newspapers Waste Through Alkali and Enzymatic Delignification', *Waste and Biomass Valorization*. Springer Netherlands. doi: 10.1007/s12649-017-9871-7.

**Box, G. E. P., Hunter, W. G. and Hunter, J. S.** (1978) *Statistics for Experimenters.* New York, USA: John Wiley and Sons.

**Brooks, T. A. and Ingram, L. O.** (1995) 'Conversion of Mixed Waste Office Paper to Ethanol by Genetically Engineered Klebsiella oxytoca Strain P2', *Biotechnology Progress*, 11(6), pp. 619–625. doi:

10.1021/bp00036a003.

**Byadgi, S. A. and Kalburgi, P. B.** (2016) 'Production of Bioethanol from Waste Newspaper', *Procedia Environmental Sciences*. The Author(s), 35, pp. 555–562. doi: 10.1016/j.proenv.2016.07.040.

**Chakraborty, A. and Mahajan, A.** (2014) 'Cellulase Activity Enhancement of Bacteria Isolated From Oil-Pump Soil Using Substrate and Medium Optimization', *American Journal of Microbiological Research*, 2(2), pp. 52–56. doi: 10.12691/ajmr-2-2-1.

**Chandel, A. K.** *et al.* (2007) 'Economics and environmental impact of bioethanol production technologies: an appraisal', *Biotechnology and Molecular Biology Review*, 2(February), pp. 14–32. Available at: http://www.academicjournals.org/BMBR.

**Chiaramonti, D.** *et al.* (2012) 'Review of pretreatment processes for lignocellulosic ethanol production, and development of an innovative method', *Biomass and Bioenergy*. Elsevier Ltd, 46, pp. 25–35. doi: 10.1016/j.biombioe.2012.04.020.

**Crespo, C. F.** *et al.* (2012) 'Ethanol production by continuous fermentation of d-(+)-cellobiose, d-(+)-xylose and sugarcane bagasse hydrolysate using the thermoanaerobe Caloramator boliviensis', *Bioresource Technology*. Elsevier Ltd, 103(1), pp. 186–191. doi: 10.1016/j.biortech.2011.10.020.

**Cullimore, D. R.** (2010) *Practical Atlas for Bacterial Identification*. 2nd Editio. USA: CRC Press.

**Dahnum, D.** *et al.* (2015) 'Comparison of SHF and SSF processes using enzyme and dry yeast for optimization of bioethanol production from empty fruit bunch', in *Energy Procedia*. Elsevier B.V., pp. 107–116. doi: 10.1016/j.egypro.2015.03.238.

**Dale, M. C. and Musgrove, D.** (2009) 'Continuous Conversion of MSW-derived Waste Paper to Bio-Ethanol Using a 100L 6-stage Continuous Stirred Reactor Separator', *Office*.

**Demirbas, A.** (2017) 'Tomorrow's biofuels: Goals and hopes', *Energy Sources, Part A: Recovery, Utilization, and Environmental Effects*, 39(7), pp. 673–679. doi: 10.1080/15567036.2016.1252815.

**Dubey, A. K.** *et al.* (2012) 'Bioethanol production from waste paper acid pretreated hydrolyzate with xylose fermenting Pichia stipitis', *Carbohydrate Polymers*. Elsevier Ltd., 88(3), pp. 825–829. doi: 10.1016/j.carbpol.2012.01.004.

**Elgharbawy, A. A. *et al.*** (2016) *Ionic liquid pretreatment as emerging approaches for enhanced enzymatic hydrolysis of lignocellulosic biomass*, *Biochemical Engineering Journal*. Elsevier B.V. doi: 10.1016/j.bej.2016.01.021.

**Fernandez-Bolanos, J. *et al.*** (2001) 'Steam-explosion of olive stones: Hemicellulose solubilization and enhancement of enzymatic hydrolysis of cellulose', *Bioresource Technology*, 79(1), pp. 53–61. doi: 10.1016/S0960-8524(01)00015-3.

**Foyle, T., Jenninggs, L. and Mulcahy, P.** (2007) 'Compositional analysis of lignocellulosic materials: Evaluation of methods used for sugar analysis of waste paper and straw', *Bioresource Technology*, 98(16), pp. 3026–3036. doi: 10.1016/j.biortech.2006.10.013.

**Galbe, M. and Zacchi, G.** (2002) 'A review of the production of ethanol from softwood', *Applied Microbiology and Biotechnology*, 59(6), pp. 618–628. doi: 10.1007/s00253-002-1058-9.

**Gámez, S. *et al.*** (2006) 'Study of the hydrolysis of sugar cane bagasse using phosphoric acid', *Journal of Food Engineering*, 74(1), pp. 78–88. doi: 10.1016/j.jfoodeng.2005.02.005.

**Gírio, F. M. *et al.*** (2010) 'Hemicelluloses for fuel ethanol: A review', *Bioresource Technology*. Elsevier Ltd, 101(13), pp. 4775–4800. doi: 10.1016/j.biortech.2010.01.088.

**Glazer, A. N. and Nikaido, H.** (2007) *MICROBIAL BIOTECHNOLOGY*. 2nd Editio. Edited by A. N. and H. N. Glazer. CAMBRIDGE UNIVERSITY PRESS.

**Granda, C. B., Zhu, L. and Holtzapple, M. T.** (2007) 'Sustainable Liquid Biofuels and Their Environmental Impact', *Environmental Progress*, 26(3), pp. 233–250. doi: 10.1002/ep.

**Guerfali, M. *et al.*** (2014) 'Enhanced Enzymatic Hydrolysis of Waste Paper for Ethanol Production Using Separate Saccharification and Fermentation', *Appl Biochem Biotechnol*. doi: 10.1007/s12010-014-1243-1.

**Haan, R. den** *et al.* (2013) 'Engineering Saccharomyces cerevisiae for next generation ethanol production', *Journal of Chemical Technology and Biotechnology*, 88(6), pp. 983–991. doi: 10.1002/jctb.4068.

**Hamelinck, C. N., Hooijdonk, G. van and Faaij, A.** (2005) 'Ethanol from lignocellulosic biomass: Techno-economic performance in short-, middle- and long-term', *Biomass and Bioenergy*, 28(4), pp. 384–410. doi: 10.1016/j.biombioe.2004.09.002.

**Haq, F.** *et al.* (2016) 'Recent progress in bioethanol production from lignocellulosic materials: A review', *International Journal of Green Energy*, 13(14), pp. 1413–1441. doi: 10.1080/15435075.2015.1088855.

**Harmsen, P. F. H.** *et al.* (2010) 'Literature Review of Physical and Chemical Pretreatment Processes for Lignocellulosic Biomass', *Wageningen University & Research centre - Food & Biobased Research*, (September), pp. 1–49.

**Imran, M.** *et al.* (2016) 'Cellulase Production from Species of Fungi and Bacteria from Agricultural Wastes and Its Utilization in Industry: A Review', *Advances in Enzyme Research*, 4(2), pp. 44–55. doi: 10.4236/aer.2016.42005.

International Energy Agency (2015) 'World Energy Outlook 2015 Factsheet', *Global energy trends to 2040:The energy sector and climate change in the run-up to COP21*, pp. 1–4. doi: 10.1787/20725302.

**Irfan, M.** *et al.* (2012) 'Isolation and screening of cellulolytic bacteria from soil and optimization of cellulase production and activity', *Turkish Journal of Biochemistry*, 37(3), pp. 287–293. doi: 10.5505/tjb.2012.09709.

**Jones, B. W.** *et al.* (2013) 'Enhancement in enzymatic hydrolysis by mechanical refining for pretreated hardwood lignocellulosics', *Bioresource Technology*. Elsevier Ltd, 147, pp. 353–360. doi: 10.1016/j.biortech.2013.08.030.

**Jonsson, L. J. and Martin, C.** (2016) 'Pretreatment of lignocellulose: Formation of inhibitory by-products and strategies for minimizing their effects', *Bioresource Technology*, 199, pp. 103–112. doi: 10.1016/j.biortech.2015.10.009.

**Jørgensen, H.** *et al.* (2003) 'Purification and characterization of five cellulases and one xylanase from Penicillium brasilianum IBT 20888', *Enzyme and Microbial Technology*, 32(7), pp. 851–861. doi: 10.1016/S0141-0229(03)00056-5.

**Kádár, Z., Szengyel, Z. and Reczey, K.** (2004) 'Simultaneous saccharification and fermentation (SSF) of industrial wastes for the production of ethanol', *Industrial Crops and Products*, 20(1), pp. 103–110. doi: 10.1016/j.indcrop.2003.12.015.

**Kafuku, G. and Mbarawa, M.** (2010) 'Alkaline catalyzed biodiesel production from moringa oleifera oil with optimized production parameters', *Applied Energy*. Elsevier Ltd, 87(8), pp. 2561–2565. doi: 10.1016/j.apenergy.2010.02.026.

**Kang, L., Wang, W. and Lee, Y. Y.** (2010) 'Bioconversion of kraft paper mill sludges to ethanol by SSF and SSCF', *Applied Biochemistry and Biotechnology*, 161(1–8), pp. 53–66. doi: 10.1007/s12010-009-8893-4.

**Kang, Q.** *et al.* (2014) 'Bioethanol from lignocellulosic biomass: Current findings determine research priorities', *Scientific World Journal*, 2014(Ci), pp. 1–13. doi: 10.1155/2014/298153.

**Keller, F. A., Hamilton, J. E. and Nguyen, Q. A.** (2003) 'Microbial pretreatment of biomass', *Applied biochemistry and Biotechnology*, 105–108(3), pp. 27–41. doi: 10.1385/ABAB:105:1-3:27.

**Klinke, H. B., Thomsen, A. B. and Asring, B. K.** (2004) 'Inhibition of ethanol-producing yeast and bacteria by degradation products produced during pre-treatment of biomass', *Applied Microbiology and Biotechnology*, 66(1), pp. 10–26. doi: 10.1007/s00253-004-1642-2.

**Krishna, B. S. and Kakadiya,** U. K. (2015) 'Performance evaluation for Bioethanol production in three reactors : Batch , packed bed and Fluidized bed bio-reactors', *International Journal on Applied Bioengineering*, 9(1).

**Kuhad, R. C.** *et al.* (2010) 'Fed batch enzymatic saccharification of newspaper cellulosics improves the sugar content in the hydrolysates and eventually the ethanol fermentation by Saccharomyces cerevisiae', *Biomass and Bioenergy*. Elsevier Ltd, 34(8), pp. 1189–1194. doi: 10.1016/j.biombioe.2010.03.009.

**Kuhad, R., Gupta, R. and Singh, A.** (2011) 'Microbial Cellulases and

Their Industrial Applications', *Enzyme Research*, 2011, pp. 1–10. doi: 10.4061/2011/280696.

**Kumar, P. *et al.*** (2009) 'Methods for pretreatment of lignocellulosic biomass for efficient hydrolysis and biofuel production', *Industrial and Engineering Chemistry Research*, pp. 3713–3729. doi: 10.1021/ie801542g.

**Lee, D. H. *et al.*** (2010) 'Pretreatment of waste newspaper using ethylene glycol for bioethanol production', *Biotechnology and Bioprocess Engineering*, 15(6), pp. 1094–1101. doi: 10.1007/s12257-010-0158-0.

**Lee, J.** (1997) 'Biological conversion of lignocellulosic biomass to ethanol - Review article', *Journal of Biotechnology*, 56, pp. 1–24.

**Lima, D. A. *et al.*** (2015) 'Comparison of bioethanol production from acid hydrolysates of waste office paper using saccharomyces', *Cellulose Chemistry and Technologyhemistry and Technology*, 49(5–6), pp. 463–469.

**Lin, Y. and Tanaka, S.** (2006) 'Ethanol fermentation from biomass resources: Current state and prospects', *Applied Microbiology and Biotechnology*, 69(6), pp. 627–642. doi: 10.1007/s00253-005-0229-x.

**Lu, J. *et al.*** (2013) 'Fed-batch semi-simultaneous saccharification and fermentation of reed pretreated with liquid hot water for bio-ethanol production using Saccharomyces cerevisiae', *Bioresource Technology*. Elsevier Ltd, 144, pp. 539–547. doi: 10.1016/j.biortech.2013.07.007.

**Lu, L. and Zhou, J.** (2014) 'A review of environmental assessments on liquid biofuels in China', *Energy Procedia*, 61, pp. 2105–2108. doi: 10.1016/j.egypro.2014.12.086.

**Luzzi, S. C. *et al.*** (2017) 'Pretreatment of lignocellulosic biomass using ultrasound aiming at obtaining fermentable sugar', *Biocatalysis and Biotransformation*. Informa UK Limited, trading as Taylor 8 Francis Group, pp. 1–7. doi: 10.1080/10242422.2017.1310206.

**Lynd, L. R.** (1996) 'Overview and evaluation of fuel ethanol from cellulosic biomass: Technology, economics, the environment, and policy', *Annual Review of Energy and the Environment*, 21, pp. 403–465. doi: 10.1146/annurev.energy.21.1.403.

**Lynd, L. R. *et al.*** (2002) 'Microbial Cellulose Utilization: Fundamentals and Biotechnology Lee', *MICROBIOLOGY AND MOLECULAR BIOLOGY*

*REVIEWS*, 66(3), pp. 506–577. doi: 10.1128/MMBR.66.3.

**Lynd, L. R.** *et al.* (2005) 'Consolidated bioprocessing of cellulosic biomass: An update', *Current Opinion in Biotechnology*, 16(5), pp. 577–583. doi: 10.1016/j.copbio.2005.08.009.

**Maceiras, R. and Alfonsín, V.** (2017) 'Bioethanol Production from Waste Office Paper', *Journal of Environmental Science*, 6(1), pp. 45–49.

**Maki, M., Leung, K. T. and Qin, W.** (2009) 'The prospects of cellulase-producing bacteria for the bioconversion of lignocellulosic biomass', *International Journal of Biological Sciences*, 5(5), pp. 500–516. doi: 10.7150/ijbs.5.500.

**Mariam, I.** *et al.* (2009) 'Enhanced Production of Ethanol From Free and Under Stationary Culture', *Mutation Research/Fundamental and Molecular Mechanisms of Mutagenesis*, 41(2), pp. 821–833. doi: 10.1016/0027-5107(69)90103-1.

**Martín, C., Klinke, H. B. and Thomsen, A. B.** (2007) 'Wet oxidation as a pretreatment method for enhancing the enzymatic convertibility of sugarcane bagasse', *Enzyme and Microbial Technology*, 40(3), pp. 426–432. doi: 10.1016/j.enzmictec.2006.07.015.

**Matsushika, A.** *et al.* (2009) 'Ethanol production from xylose in engineered Saccharomyces cerevisiae strains: Current state and perspectives', *Applied Microbiology and Biotechnology*, 84(1), pp. 37–53. doi: 10.1007/s00253-009-2101-x.

**Mawadza, C.** *et al.* (2000) 'Purification and characterization of cellulases produced by two Bacillus strains.', *Journal of biotechnology*, 83(3), pp. 177–187. Available at: http://www.sciencedirect.com/science/article/pii/S0168165600003059.

**Menendez, E., Fraile, P. G. and Rivas, R.** (2015) 'Biotechnological applications of bacterial cellulases', *AIMS Bioengineering*, 2(3), pp. 163–182. doi: 10.3934/bioeng.2015.3.163.

**Miller, G. L.** (1959) 'Use of Dinitrosalicylic Acid Reagent for Determination of Reducing Sugar', *ANALYTICAL CHEMISTRY*, 31(3), pp. 426–428.

**Montgomery, D.** (2013) *Design and Analysis of Experiments*. North

Carolina, USA: SAS Institute Inc.

**Mood, S. H.** *et al.* (2013) 'Lignocellulosic biomass to bioethanol, a comprehensive review with a focus on pretreatment', *Renewable and Sustainable Energy Reviews*, 27, pp. 77–93. doi: 10.1016/j.rser.2013.06.033.

**Mosier, N.** *et al.* (2005) 'Features of promising technologies for pretreatment of lignocellulosic biomass', *Bioresource Technology*, 96(6), pp. 673–686. doi: 10.1016/j.biortech.2004.06.025.

**Muthsamy, S.** *et al.* (2017) 'Bioconversion and bioethanol production from agro-residues through fermentation process using mangrove-associated actinobacterium Streptomyces olivaceus (MSU3)', *Biofuels*. Taylor & Francis, 7269(April), pp. 1–13. doi: 10.1080/17597269.2017.1309853.

**Nair, R. B., Lennartsson, P. R. and Taherzadeh, M. J.** (2016) 'Bioethanol Production From Agricultural and Municipal Wastes', in *Current Developments in Biotechnology and Bioengineering*. Elsevier B.V., pp. 157–190. doi: 10.1016/B978-0-444-63664-5.00008-3.

**Najafpour, G. D.** (2015) *BIOCHEMICAL ENGINEERING AND BIOTECHNOLOGY*. 2nd Edition. Elsevier.

**Nandimath, A. P.** *et al.* (2016) 'Optimization of cellulase production for Bacillus sp. and Pseudomonas sp. soil isolates', *African Journal of Microbiology Research*, 10(13), pp. 410–419. doi: 10.5897/AJMR2016.7954.

**Nishimura, H.** *et al.* (2017) 'Production of ethanol from a mixture of waste paper and kitchen waste via a process of successive liquefaction , presaccharification , and simultaneous saccharification and fermentation', *Waste Management*, 66, pp. 226–235. doi: 10.1016/j.wasman.2017.04.030.

**Niven, R. K.** (2005) 'Ethanol in gasoline: Environmental impacts and sustainability review article', *Renewable and Sustainable Energy Reviews*, 9(6), pp. 535–555. doi: 10.1016/j.rser.2004.06.003.

**Öhgren, K.** *et al.* (2007) 'A comparison between simultaneous saccharification and fermentation and separate hydrolysis and fermentation using steam-pretreated corn stover', *Process Biochemistry*, 42(5), pp. 834–839. doi: 10.1016/j.procbio.2007.02.003.

**Ong, L. G. A.** *et al.* (2004) 'Enzyme production and profile by Aspergillus niger during solid substrate fermentation using palm kernel cake as substrate.', *Applied biochemistry and biotechnology*, 118, pp. 73–79. doi: 10.1385/ABAB:118:1-3:073.

**Otajevwo, F. D. and Aluyi, H. S. A.** (2011) 'Cultural conditions necessary for optimal cellulase yield by cellulolytic bacterial organisms as they relate to residual sugars released in broth medium', *Modern Applied Science*, 5(3), pp. 141–151. doi: 10.5539/mas.v5n3p141.

**Pan, X.** *et al.* (2006) 'Bioconversion of Hybrid Poplar to Ethanol and Co-Products Using an Organosolv Fractionation Process: Optimization of Process Yields', *Biotechnology and Bioengineering*, 94(5), pp. 851–861. doi: 10.1002/bit.20905.

**Park, N.** *et al.* (2010) 'Organosolv pretreatment with various catalysts for enhancing enzymatic hydrolysis of pitch pine (Pinus rigida)', *Bioresource Technology*, 101(18), pp. 7046–7053. doi: 10.1016/j.biortech.2010.04.020.

**Pathak, V. M. and Navneet, A. S.** (2016) 'Screening and Characterization of Cellulose Degrading Bacterial Isolates of Waste disposal Site', *International Journal of Current Microbiology and Applied Sciences*, 5(6), pp. 898–907. doi: http://dx.doi.org/10.20546/ijcmas.2016.506.097.

**Paulová, L.** *et al.* (2013) 'Chapter 2, Production of 2nd Generation of Liquid Biofuels', in Fang, Z. (ed.) *Liquid, Gaseous and Solid Biofuels - Conversion Techniques.* InTech, pp. 47–78. doi: http://dx.doi.org/10.5772/53492.

**Poulsen, O. M. and Petersen, L. W.** (1988) 'Growth of Cellulomonas sp. ATCC 21399 on different polysaccharides as sole carbon source Induction of extracellular enzymes', *Appl Microbiol Biotechnol*, 29, pp. 480–484.

**Prasad, S., Singh, A. and Joshi, H. C.** (2007) 'Ethanol as an alternative fuel from agricultural, industrial and urban residues', *Resources, Conservation and Recycling*, 50(1), pp. 1–39. doi: 10.1016/j.resconrec.2006.05.007.

**Prema, D., Prabha, M. L. and Gnanavel, G.** (2015) 'Production of Biofuel using Waste Papers from Pseudomonas aeruginosa', *International Journal of ChemTech Research*, 8(4), pp. 974–4290.

**Rajasekaran, R., Vijayaraghavan, G. and Marimuthu, C.** (2014)

'Synthesis of bio-ethanol by saccharomyces cerevisiae using lignocellulosic hydrolyzate from pretreated waste paper fermentation', *International Journal of ChemTech Research*, 6(11), pp. 4904–4912.

**Rajoka, M. and Malik, K. A.** (1997) 'CELLULASE PRODUCTION BY CELLULOMONAS BIAZOTEA CULTURED IN MEDIA CONTAINING DIFFERENT CELLULOSIC SUBSTRATES', *Bioresource Technology*, 59(1), pp. 21–27. doi: 10.1016/S0960-8524(96)00136-8.

**Rocha, G. J. *et al.*** (2011) 'Dilute mixed-acid pretreatment of sugarcane bagasse for ethanol production', *Biomass and Bioenergy*. Elsevier Ltd, 35(1), pp. 663–670. doi: 10.1016/j.biombioe.2010.10.018.

**Rocha, J. M.** (2016) 'Enzymatic hydrolysis of waste office paper for ethanol production by Spathaspora Passalidarum', *CELLULOSE CHEMISTRY AND TECHNOLOGY*, 50(2), pp. 243–246.

**Rosenberger, A. *et al.*** (2002) 'Costs of bioethanol production from winter cereals: The effect of growing conditions and crop production intensity levels', *Industrial Crops and Products*, 15(2), pp. 91–102. doi: 10.1016/S0926-6690(01)00099-1.

**Sadhu, S. and Maiti, T. K.** (2013) 'Cellulase Production by Bacteria : A Review', *British Microbiology Research Journal*, 3(3), pp. 235–258. doi: 10.9734/BMRJ/2013/2367.

**Saggi, S. K. and Dey, P.** (2016) 'An overview of simultaneous saccharification and fermentation of starchy and lignocellulosic biomass for bio-ethanol production', *Biofuels*, 7269(September), pp. 1–13. doi: 10.1080/17597269.2016.1193837.

**Saha, B. C. and Cotta, M. A.** (2007) 'Enzymatic saccharification and fermentation of alkaline peroxide pretreated rice hulls to ethanol', *Enzyme and Microbial Technology*, 41(4), pp. 528–532. doi: 10.1016/j.enzmictec.2007.04.006.

**Saini, J. K.** (2015) 'Lignocellulosic agriculture wastes as biomass feedstocks for second-generation bioethanol production: concepts and recent developments', *3 Biotech*, pp. 337–353. doi: 10.1007/s13205-014-0246-5.

**Sangkharak, K.** (2011) 'Optimization of enzymatic hydrolysis for ethanol production by simultaneous saccharification and fermentation of wastepaper', *Waste Management & Research*, 29(11), pp. 1134–1144. doi: 10.1177/0734242X10387656.

**Sangkharak, K., Vangsirikul, P. and Janthachat, S.** (2011) 'Isolation of Novel Cellulase From Agricultural Soil and Application for Ethanol Production', *International Journal of Advanced Biotechnology and Research*, 2(2), pp. 230–239. Available at: http://www.bipublication.com.

**Sarkar, N. *et al.*** (2012) 'Bioethanol production from agricultural wastes: An overview', *Renewable Energy*. Elsevier Ltd, 37(1), pp. 19–27. doi: 10.1016/j.renene.2011.06.045.

**Schulein, M.** (1997) 'Enzymatic properties of cellulases from Humicola insolens', *Journal of Biotechnology*, 57, pp. 71–81.

**Sethi, S. *et al.*** (2013) 'Optimization of Cellulase Production from Bacteria Isolated from Soil', *ISRN Biotechnology*, 2013, pp. 1–7. doi: 10.5402/2013/985685.

**Sharada, R. *et al.*** (2014) 'Applications of Cellulases – Review', *International Journal of Pharmaceutical, Chemical and Biological Sciences*, 4(2), pp. 424–437.

**Sheehan, J. *et al.*** (2004) 'Energy and Environmental Aspects of Using Corn Stover for Fuel Ethanol', *Journal of Industrial Ecology*, 7(3), pp. 117–146. doi: 10.1162/108819803323059433.

**Shi, J. *et al.*** (2009) *The Potential of Cellulosic Ethanol Production from Municipal solid waste: A Technical and Economic Evaluation, Energy Development and Technology 015.* Available at: http://citeseerx.ist.psu.edu/viewdoc/download?doi=10.1.1.524.7444&rep=rep1&type=pdf.

**Silverstein, R. A. *et al.*** (2007) 'A comparison of chemical pretreatment methods for improving saccharification of cotton stalks', *Bioresource Technology*, 98(16), pp. 3000–3011. doi: 10.1016/j.biortech.2006.10.022.

**Singh, D. P. and Trivedi, R. K.** (2014) 'Biofuel from wastes an economic and environmentally feasible resource', *Energy Procedia*. Elsevier B.V., 54, pp. 634–641. doi: 10.1016/j.egypro.2014.07.305.

**Sivers, M. V. and Zacchi, G.** (1995) 'A techno-economical comparison of three processes for the production of ethanol from pine', *Bioresour.*

*Technol.*, 51, pp. 43–52.

**Sluiter, A.** *et al.* (2008) *Determination of Ash in Biomass: Laboratory Analytical Procedure (LAP), National Renewable Energy Laboratory (NREL).*

**Sluiter, A.** *et al.* (2012) *Determination of Structural Carbohydrates and Lignin in Biomass: Laboratory Analytical Procedure (LAP), National Renewable Energy Laboratory (NREL).*

**Solomon, B. D., Barnes, J. R. and Halvorsen, K. E.** (2007) 'Grain and cellulosic ethanol: History, economics, and energy policy', *Biomass and Bioenergy*, 31(6), pp. 416–425. doi: 10.1016/j.biombioe.2007.01.023.

**Srivastava, N.** *et al.* (2015) 'A Review on Fuel Ethanol Production From Lignocellulosic Biomass', *International Journal of Green Energy*, 12(9), pp. 949–960. doi: 10.1080/15435075.2014.890104.

**Subhedar, P. B. and Gogate, P. R.** (2015) 'Ultrasound-assisted bioethanol production from waste newspaper', *Ultrasonics Sonochemistry*, 27, pp. 37–45. doi: 10.1016/j.ultsonch.2015.04.035.

**Sukumaran, R. K., Singhania, R. R. and Pandey, A.** (2005) 'Microbial cellulases - Production, applications and challenges', *Journal of Scientific and Industrial Research*, 64(11), pp. 832–844.

**Sun, Y. and Cheng, J.** (2002) 'Hydrolysis of lignocellulosic materials for ethanol production', *Bioresource Technology*, 83(83), pp. 1–11.

**Taha, M.** *et al.* (2016) 'Commercial feasibility of lignocellulose biodegradation: Possibilities and challenges', *Current Opinion in Biotechnology*. Elsevier Ltd, 38, pp. 190–197. doi: 10.1016/j.copbio.2016.02.012.

**Taherzadeh, M. J. and Karimi, K.** (2007a) 'Acid-based hydrolysis processes for ethanol from lignocellulosic materials: A review', *BioResources*, 2(3), pp. 472–499. doi: 10.15376/biores.2.3.472-499.

**Taherzadeh, M. J. and Karimi, K.** (2007b) 'Enzyme-based hydrolysis processes for ethanol from lignocellulosic materials: A review', *BioResources*, 2(4), pp. 707–738. doi: 10.15376/BIORES.2.3.472-499.

**Tan, H. T., Lee, K. T. and Mohamed, A. R.** (2011) 'Pretreatment of lignocellulosic palm biomass using a solvent-ionic liquid [BMIM]Cl for glucose recovery: An optimisation study using response surface methodology', *Carbohydrate Polymers*. Elsevier Ltd., 83(4), pp. 1862–1868. doi: 10.1016/j.carbpol.2010.10.052.

**Travaini, R.** *et al.* (2013) 'Sugarcane bagasse ozonolysis pretreatment: Effect on enzymatic digestibility and inhibitory compound formation', *Bioresource Technology*, 133, pp. 332–339. doi: 10.1016/j.biortech.2013.01.133.

**Varotkar, P. V., Tumane, P. M. and Wasnik, D. D.** (2016) 'Bioconversion of Waste Paper into Bio-Ethanol by Co-Culture of Fungi Isolated from Lignocellulosic Waste', *International Journal of Pure and Applied Bioscience*, 4(4), pp. 264–274. doi: 10.18782/2320-7051.2329.

**Wackett, L. P.** (2008) 'Biomass to fuels via microbial transformations', *Current Opinion in Chemical Biology*, 12(2), pp. 187–193. doi: 10.1016/j.cbpa.2008.01.025.

**Wang, L.** *et al.* (2012a) 'Bioethanol production from various waste papers: Economic feasibility and sensitivity analysis', *Applied Energy*. Elsevier Ltd, 111, pp. 1172–1182. doi: 10.1016/j.apenergy.2012.08.048.

**Wang, L.** *et al.* (2012b) 'Technology performance and economic feasibility of bioethanol production from various waste papers', *Energy Environ. Sci.*, 5(2), pp. 5717–5730. doi: 10.1039/C2EE02935A.

**Wayman, M., Chen, S. and Doan, K.** (1992) 'Bioconversion of waste paper to ethanol', *Process Biochemistry*, 27(4), pp. 239–245. doi: 10.1016/0032-9592(92)80024-W.

**Wu, B.** *et al.* (2001) 'Factors Controlling Alkylbenzene Sorption to Municipal Solid Waste', *ENVIRONMENTAL SCIENCE & TECHNOLOGY*, 35, pp. 4569–4576.

**Wu, F. C., Huang, S. S. and Shih, I. L.** (2014) 'Sequential hydrolysis of waste newspaper and bioethanol production from the hydrolysate', *BIORESOURCE TECHNOLOGY*. Elsevier Ltd, 167, pp. 159–168. doi: 10.1016/j.biortech.2014.06.041.

**Wu, M., Wang, M. and Huo, H.** (2006) *Fuel-Cycle Assessment of Selected Bioethanol Production Pathways in the United States*, *Center for*

*Transportation Research Energy Systems Division, Argonne National Laboratory*. doi: ANL/ESD/06-7.

**Yang, B.** *et al.* (2011) 'Enzymatic hydrolysis of cellulosic biomass', *Biofuels*, 2(4), pp. 421–449. doi: 10.4155/bfs.11.116.

**Yang, B. and Wyman, C. E.** (2004) 'Effect of Xylan and Lignin Removal by Batch and Flowthrough Pretreatment on the Enzymatic Digestibility of Corn Stover Cellulose', *Biotechnology and Bioengineering*, 86(1), pp. 88–95. doi: 10.1002/bit.20043.

**Yin, L. J., Huang, P. S. and Lin, H. H.** (2010) 'Isolation of cellulase-producing bacteria and characterization of the cellulase from the isolated bacterium cellulomonas Sp. YJ5', *Journal of Agricultural and Food Chemistry*, 58(17), pp. 9833–9837. doi: 10.1021/jf1019104.

**Zabed, H.** *et al.* (2016) 'Fuel ethanol production from lignocellulosic biomass: An overview on feedstocks and technological approaches', *Renewable and Sustainable Energy Reviews*. Elsevier, 66, pp. 751–774. doi: 10.1016/j.rser.2016.08.038.

**Zainab, B. and Fakhra, A.** (2014) 'Production of Ethanol by fermentation process by using Yeast Saccharomyces cerevisae', *International Research Journal of Environment Sciences*, 3(7), pp. 24–32.

**Zhang, J. and Lynd, L. R.** (2010) 'Ethanol production from paper sludge by simultaneous saccharification and co-fermentation using recombinant xylose-fermenting microorganisms', *Biotechnology and Bioengineering*, 107(2), pp. 235–244. doi: 10.1002/bit.22811.

**Zhang, W.** *et al.* (2013) 'Optimisation of simultaneous saccharification and fermentation of wheat straw for ethanol production', *Fuel*. Elsevier Ltd, 112, pp. 331–337. doi: 10.1016/j.fuel.2013.05.064.

**Zhang, X. Z. and Zhang, Y. H. P.** (2013) *CELLULASES: Characteristics, Sources, Production , and Applications*. 1st Editio, *Bioprocessing Technologies in Biorefi nery for Sustainable Production of Fuels*. 1st Editio. Edited by S.-T. Yang. John Wiley & Sons, Inc. doi: 10.1002/9781118642047.ch8.

**Zhang, Y. H. P., Himmel, M. E. and Mielenz, J. R.** (2006) 'Outlook for cellulase improvement: Screening and selection strategies', *Biotechnology*

*Advances*, 24(5), pp. 452–481. doi: 10.1016/j.biotechadv.2006.03.003.

**Zhao, X., Cheng, K. and Liu, D.** (2009) 'Organosolv pretreatment of lignocellulosic biomass for enzymatic hydrolysis', *Applied Microbiology and Biotechnology*, 82(5), pp. 815–827. doi: 10.1007/s00253-009-1883-1.

**Zhao, X. Q.** *et al.* (2012) 'Bioethanol from Lignocellulosic Biomass', *Advances in biochemical engineering/biotechnology*, 128, pp. 25–51. doi: 10.1007/10_2011_129.

**Zichová, M.** *et al.* (2017) 'Production of ethanol from waste paper using immobilized yeasts', *Chemical Papers*, 71(3). doi: 10.1007/s11696-016-0036-0.

**Zyl, W. H. V.** *et al.* (2007) 'Consolidated bioprocessing for bioethanol production using saccharomyces cerevisiae', *Advances in Biochemical Engineering/Biotechnology*, 108(April), pp. 205–235. doi: 10.1007/10_2007_061.

# Table of contents

More
Books!

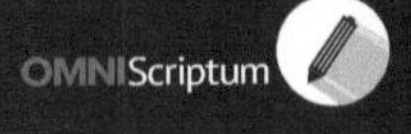

OMNIScriptum

FSC
www.fsc.org
MIX
Papier aus verantwortungsvollen Quellen
Paper from responsible sources
FSC® C105338